imagine 08 – CONCRETABLE

Delft University of Technology, Faculty of Architecture,
Chair of Design of Construction

imagine 08

SERIES EDITED BY
Ulrich Knaack
Tillmann Klein
Marcel Bilow

CONCRETABLE

Ulrich Knaack
Sascha Hickert
Linda Hildebrand

nai010 publishers, Rotterdam 2015

CONTENTS

CONCRETE – MORE CAPABLE THAN YOU THINK

Over the last 150 years concrete has coined the building industry and has proven itself as a predestined material with a broad range of application. Due to its longevity laymen often conceive this material as a synonym for unimaginative backwardness and old-fashioned lack of originality. It is often perceived as 'unsexy'. But it needn't be that way. Increasingly, future-oriented architects and engineers recognize the material's enormous potential with regards to creative formability and attractive surface design. Modern manufacturing methods allow for filigree load-bearing structures and innovative cross-sections.

This volume demonstrates in an impressive manner the many facets of the material concrete, and introduces unusual approaches for sustainable buildings. It becomes clear how, with concrete, we can expand the architectural design vocabulary and align it with today's technical requirements. The authors document how creative collaboration and unconventional thought experiments between architects thinking outside the box and imaginative engineers can produce entirely new and surprising ideas for the use of the building material concrete. Herewith, concrete proves its transformation ability and great potential for innovation.

Univ.-Prof. Dr.-Ing. Carl-Alexander Graubner
Institute for Solid Construction
Technical University Darmstadt

1. INTRODUCTION

WHAT

Concrete is massive, gray and cold! It is usually associated with brutal, massive and depressing architecture – we all have similar pictures in our mind. And when talking to concrete technology specialists all we hear is "everything is investigated – we know it all". Too bad – so it's a material at its final destination: no new developments, no innovation, no inspiring and thrilling future. Well yes, there are some positive features – great load-bearing capacity, high fire resistance, large thermal mass. But alas, The End!

However, there have been several developments over the last decades: ultra-high performance concrete, with performance capability close to that of steel – while being able to be very easily molded into a multitude of shapes. Or self-compacting concrete, which allows increasing reinforcement and simultaneously provides better controlled surface qualities. And the combination with stainless steel or glass fiber reinforcements, which allow thinner or even ductile constructions. Finally, investigations in the area of embodied energy and carbon dioxide storage in concrete, energized by sunlight.

So – is there an exciting future after all?

When looking at the material developments it becomes obvious that with a material, which can take forces like steel and that can, to a certain extent, become ductile, the construction we aim for is no longer necessarily massive but could become light – light in the sense of its weight due to the limited load and light in the sense of its appearance: the material is now able to concentrate forces in one point like the typical skeleton constructions, but at the same time it does not have to be connected by prefabricated components (bolts, screws etc). So the new concrete is precise, more efficient and allows more control of the aesthetic – how nice. So an all-rounder as a material – with the one drawback of involving an exothermal process with no option to convert the result.

WHY

Being driven by the interest of building things, developing things – wanting to feel them, touch them and being able to understand their potential – this imagine book focuses on concrete; or more concretely – a new understanding of concrete: light, emotional and providing more functionalities then "just" massivity and load-bearing capacity.

On one hand this is done to provide designers and engineers with potential technologies for future projects, and to develop the technologies themselves on the other. In this regard we believe that experimenting is the key instrument to ensure technical possibilities – as well as aesthetical and functional results.

Projects are conducted with an open minded motivation – to find technical driven design results for future inspiration.

WHO

The background of this activity is work done by the Facade Research Group, a research group founded at TU Delft / Faculty of Architecture (The Netherlands) and being linked to the Hochschule Ostwestfalen Lippe / Detmolder Schule für Architektur und Innenarchitektur (Germany). The group is interested in new technologies, performance development and innovation, being tested in designs and experimental buildings. Following a roadmap of possible future façade development, the group organizes itself via regular sessions, symposia and workshops, sharing knowledge and experience. There are between ten and fifteen actively pursued PhD studies, investigations in the fields of "Climate, Comfort and Energy", "Construction, Product and Material", Production and Assembly" and "Design Tools and Strategies", financed via various resources ranging from industry research, national and international grants as well as individual interests.

The research presented with this imagine book is part of the research field "Production and Assembly" and is partly financed by a grant of the Deutsche Forschungsgemeinschaft (DFG), the scientific German grant agency for fundamental research for the project. It is part of the Schwerpunktprogramm (SPP) 1542 "Leicht Bauen mit Beton" (special research program "building light with concrete"). Within this environment the authors collaborate closely with Prof. Dr. C. A. Graubner and his staff at the chair Massivbau / TU Darmstadt (Germany) in the area of adjustable molding technologies and fabric formwork.

Being part of the imagine series, a series of books about building envelope related technologies to inspire designers and engineers, this books offers a status quo of the technology, potentials for development and suggestions for future performances. As is typical of the series' layout, additional built projects and principles as well as concept sketches of technologies are provided. For further reading please also visit the following sites:

http://concretable.de/
http://facadeworld.com/
http://imagineblog.tumblr.com/

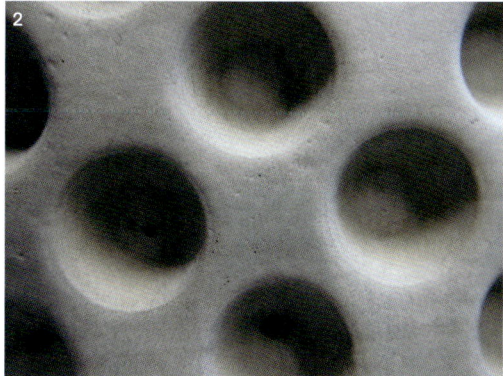

1 Wooden concrete
2 Formliner concrete
3 Standard Hotel New York

2. MATERIAL

2.1 INTRODUCTION

Innovation develops in two ways: The push operation: a new technology is put on the market, requiring some time to be accepted by early users, who take all the risk of failure but also the advantage of additional performance and glory. And the pull operation, in which a user/decision maker defines a task, not yet technically accomplished but principally realizable. Innovative players suggest solutions – and if successful, build the object. Again, the advantage of additional performance and glory.

A prominent area of innovation, partly pull and partly push driven, is material technology. Coming in waves of innovations, driven by individuals and/or more global aspects such as the need of energy reduction or aesthetical trends, materials are often promoted as initiator for innovation. The development of iron/steel are examples, being pushed by the industrialization age and the need for infrastructure, later transferred into building construction because of their effectiveness. Or the more recent developments of the material glass; mainly appreciated for being transparent but now – after decades of development – also raising interest because glass can be quite effective in collecting energy, providing insulation and even building service performances (keyword twin face facades).

And last but certainly not least the material concrete: this chapter will give an overview of the material developments of concrete from its beginnings, the key innovation of integration of stress components and potential developments for the current decade. In addition, we will focus on lightness of construction – seemingly an oxymoron but also the only real potential for the material to be used in the future when considering the global need to reduce energy. And, as engineers AND designers, we feel that the aspects of surface appearance are an underestimated area of research. Here, the focus of the contribution is on purposefully neglecting the cliché and pointing out the influencing technical factors such as formwork material, formwork method, material ingredients and finishing techniques.

Interested in the physical results, functionality and appearance, this is the place to express the strategy of experimental evolution: research delivers technical and functional results – with the consequence of the need for in-depth investigation and evaluation. The alternative – maybe unconventional but from our perspective a successful strategy – is to provoke coincidence. Thus, in order to identify potential technologies and trigger new solutions we execute experiments in a structured manner. Most of them will fail – without failing we would not have moved the boundary of the possible – but a few will be successful. And with these, further elaborated and engineered, future potential appears – and new concepts are born.

2.2 THE HISTORY OF CONCRETE

THE DEVELOPMENT OF CONCRETE'S BINDING MATERIAL

The history of concrete is closely entangled with the development of concrete formwork. The first discoveries take us back to the early Stone Age (2,000 BCE) when clay bricks where formed with wooden forms. These are considered to be the first known formwork. During the second century CE the Greeks began constructing in an entirely manner: Instead of stacking bricks and mortar they developed the technique of cast masonry (Greek: emplecton): two wall shells made of ashlar (finest stone masonry unit) or wooden boards were used as formwork, quarry stones of different size were placed in between and doused with lime mortar as binding material. This made for stable and permanent walls which are closely related to concrete constructions.

The Greeks used lime as binding material, and the Romans developed this further by changing the composition and creating the first permanent binding material. They burned lime and added powder-shaped Puzzolan (natural stone- predominantly volcanic ash), sand and crushed rocks (Marcus Vitruvius Pollio, Prestel 1959) and water which resulted in liquid stone. This was called 'Opus Caementitium' and was the first type of concrete which was also able to harden under water. It was also used with lost formwork but proved to be stronger and weather resistant. The consistence of Opus Caementitium in buildings today can be considered similar to those of approximately 2,000 year ago.

Having invented the arch construction Roman master builders combined that technique with their new concrete material. During this epoch admirable buildings were built with Opus Caementitium, for example the Coliseum and the Pantheon (both in Rome) which still exist today. But with the collapse of the Roman Empire the knowledge of Opus Caementitium vanished.

For almost 1,500 years Opus Caementitium remained unused until the transition from Renaissance to Baroque. In the 17th century alternatives for wood and clay buildings were wanted and the binding material trass was developed. This was the basis for further research on what we call cement today. As a consequence, formwork was reintroduced, and for example St. Peter's Basilica in Rome was built.

At the end of the 18th century a British engineer (John Smeaton) re-discovered a material composition which did not only harden when exposed to air but also under water. This new binding material was named 'Roman cement'. In the early 18th sand and crushed rocks were added to cement for the first time, the result being named concrete.

In the 19th century, again in Great Britain, a building contractor from the city Portland harvested clay and limestone from natural resources; burned them together and then ground the result, thus developing Portland cement for which a patent was applied in 1824. These were the beginnings of the product as we know it today.

The basis for the binding material was made 2,000 years ago. The name cement however was developed in the 19th century.

1 Pantheon Rome, view rear of the building
2 Pantheon Rome, view of the entrance
3 Pantheon Rome, View of the oculus
4 Rotunda of Mosta, Malta
5 Temple Ggantija of Gozo, Malta
6 Rotunda of Mosta, view of the dome

7 Cast masonry
8 Defensive wall of Detmold
9 Steel fibre concrete
10 Different fibres – reinforcement
11 Steel reinforcement

CONCRETE'S ABILITIES

The history of concrete in the 19th century is coined by the material combination of iron/steel and concrete, the so-called reinforced concrete. Concrete is able to assimilate high compressive forces but collapses under tensile loads. Iron and steel however cope very well when absorbing tension.

Frenchman Josepf Monier is considered to be the inventor of reinforced concrete. Since 1849, gardener and businessman Monier produced flower buckets made of concrete and bead wires. Monier discovered that those buckets were extraordinary durable. At the second world fair in Paris in 1867 Monier presented his "armed" charging bucket and registered his first patent during the same year. In a consecutive patent Monier formulated the fundamental idea to place reinforcement in those areas where the final concrete structure is subject to tensile loads. The term Monier iron (rebar steel) is still used today. And for the first time it was documented on paper that cement protects iron from rusting (Schmitt, 2001).

Already in 1874 an earlier patent was applied for which suggested to intermix metallic waste to improve the concrete's behavior but the term steel fibers was only introduced later at the end of the 1960ies.

Fiber reinforcement

The material mix has proven to solve the lacking capability of absorbing tensile forces, and now allowed for thinner profiles. Austrian industrialist Ludwig Hatschek added asbestos fibers to cementitious material and registered the patent for Eternit in 1900. This was the official beginning of fiber reinforcement concrete. It was discovered that liquid concrete weakens the fiber's capabilities. A major improvement was established in 1971: glass producer Pilkington Brothers Ltd. (England) offered glass fibers (AR= Alkaline Resistant) to the concrete industry as a material that is – unlike previous fibers – resistant against the alkaline milieu of concrete. Today, a variety of fibers is available used as reinforcement (carbon, hemp, synthetics and many more).

In the beginning of the 20th century the capabilities of concrete were improved further by the invention of high-strength concrete. This type of concrete has higher initial strength and is thus able to stay in place on a curved formwork which enhanced the possibilities of shell structures. For the German Zeiss-Planetarium, concrete producer Dyckerhoff & Widmann developed a technique to apply this product on a shell formwork (they shot concrete onto the formwork (shotcrete)). This development established the era of concrete shell structure. The most relevant representatives are Pier Luigi Nervi (since 1925), Felix Candela (since 1950), Heinz Isler (since 1950) and Ulrich Müther (since 1964).

In the building sector shotcrete decreased in relevance. Today it is most commonly used to consolidate rocks, repair and maintain concrete components.

Concrete innovation

After the Second World War concrete construction was in its peak phase. Technological progress, the availability of building rubble and especially the need for new buildings stimulated a broad range of concrete innovations. Cost and labor reduction were the main targets. The predecessors for the today so-called recycling concrete (RC concrete) was made during that time. Bending machines, concrete batching machines with huge load capacities and the introduction of system formwork technology enabled economically efficient construction.

In the beginning in the 1990ies a variety of concrete products was developed. A significant characteristic changed; whereas before concrete was made out of the three components cement, aggregates and water it developed in to a five component system (cement, aggregates, water plus powder shaped admixtures (for example fly ash) and liquid admixtures). These types of concrete exhibit a variety of new performance capabilities such as self-compression and extraordinary strength with slim profiles.

One major difference is the material's susceptibility for slight variations in ingredients; while three component concrete types can accept small variation in composition without significant changes in functionality; five component systems might actually malfunction.

Due to this extreme sensitivity to any changes in recipe, exact measurements need to be maintained and the production conditions need to be controlled. This leads to high requirements during the production process, which can only take place at the factory. The concrete is delivered to the site in liquid form. The production of five component concrete is an engineering task and requires a high level of accuracy.

Conclusion

The main innovative steps in concrete technology have been made after the Second World War, especially during the 1990ies. Major developments were the five component system and the use of modular framework. These two parameters, the

composition and the framework, are the most relevant factors for the production process and the concrete's qualities today.

Various concrete products have been developed over the last 25 years. The eleven most common types will be introduced in the following section.

Fiber concrete

In order to replace other types of concrete reinforcement, steel, glass or plastic fibers can be added to the concrete. However, since fresh concrete is highly alkaline only alkali-resistant fibers can be used. The properties of the resulting fiber concrete are mainly defined by its mechanical properties and geometry, the quantity of added fibers, their orientation, the manufacturing process and the bonding between fiber and concrete.

Textile-reinforced concrete

Similar to the afore mentioned fiber concrete, concrete can be reinforced with textiles. Typically, technical textiles such as glass or carbon fiber are used in the form of meshes, thus the name textile-reinforced concrete. The main advantage of using textile as reinforcement material lies in the fact that the reinforcement does not or minimally need to be covered – in contrast to steel reinforced concrete – since textiles do not corrode.

Self-compacting concrete (SCC)

SCC is very viscous and self-compressing. This is achieved by an increased powder content and specialized flux material. Compacting the concrete by vibration is not necessary, saving time and money for related processing steps, while the surface is clearly more appealing than that of conventional jolted concrete. Due to its aeration properties and extraordinary flow characteristics, SCC is especially suited for elements with a high share of

12 Lasered concrete
13 Concrete mixture – five component system
14 Textile-reinforced concrete
15 Self-compacting concrete
16 Glass-fibre fabrics
17 Experimental building chemistry and constructional
   engineering

reinforcement and free formed geometries. However, its susceptibility to consistence deviations is disadvantageous. For example, concerning any variation in water content. If SCC is too stiff formwork might not be filled homogenously, and If it flows to easily the structural stability decreases due to possible demixing.

UHPC (Ultra High Performance Concrete)

Ultra high performance concrete possesses an especially high structural density and thus an extremely high compressive strength of above 200 MPa (normal concrete: 25 MPa). Besides, this type of concrete is particularly impermeable and resistant against physical or chemical influences. If fibers are added to UHPC, the result is a tensile strength of 15 MPa and flexural strength of up to 45 MPa. With these exceptional properties the concrete becomes extremely brittle so that the addition of fibers is often necessary. UHPC is more consistent than standard concrete. The major advantage is its higher load-bearing at lower material thicknesses. The water / concrete ratio is reduced to a minimum (0.2) (standard concrete 0.4 to 0.7) and high-performance plasticizer (PCE) as well as admixtures like Silica-fume and fly ash are added to achieve these characteristics.

Light-weight concrete

Light-weight concrete is differentiated into 'dense light-weight concrete' and 'aggregate material porous light-weight concrete'. Light-weight concrete is particularly characterized by its low weight and low heat conductance and therefore good insulating properties, in comparison to standard concrete. The stone aggregate of dense light-weight concrete owns a clearly slighter cohesiveness than standard concrete. Thus, considerable weight savings in

this case. The rock supplements of porous light-weight concrete only get slightly encased and punctually connected to each other. An enhancement of light-weight concretes are infra-light-weight concrete or ultra-light-weight concrete which have a dry gross density of 800Kg/m max. In spite of its compressive strength, ultra-light-weight concrete can still not to be used as structural concrete today.

Ductile concrete – DUCON

DUCON is a brand of micro-reinforced high performance concrete. Originally, it was developed for an explosion and bulletproof concrete. This concrete consists of sand, self-compacting mortar and a micro-reinforcement, whereby it becomes ductile and therefore energy absorbent. The compression strength can be up to 200 MPa and the tenacity up to 20 MPa. Moreover, DUCON offers new perspectives for the architectural vocabulary as it has a realizable component thickness of > 10 mm and optional cross sections. In addition to explosion protection walls this concrete is mainly used as screed, structural support for building elements and for protection against earthquakes. DUCON can, for example, be inserted instead of screed to support ceilings.

Eco-concrete

Types of concrete that minimize the inescapable environmental pollution of its production are called eco-concrete. For this purpose there are special rules, as the composition is most commonly different to the standard concrete. The first approach is to reduce the cement share while maintaining flowability, processing time span, durability and consistency. Thereby greenhouse gases can be reduced by 30 to 70 %. Environmentally friendlier admixtures such as slag sand,

18, 20 Light-weight concrete
19 Ultra High performance concrete
21, 22 Translucent concrete
23 Gradient concrete

fly ash and pulverized limestone are (being) used to reduce the percentage of Portland cement.

Photocatalytically active cement

This is a mixture in which cement functions as binding material between concrete and photocatalysts such as TiO (modified). Thereby a reaction on the surface of the building block occurs between the photocatalyst and aerial nitrogen oxide which results in nitrite / nitrate [NO3] bound on the surface. Daylight is all that is needed for the process; no additional exposition is needed. This bears advantages especially in big cities where the NO as acid corrosive gas has reached partially unhealthy dimensions. Up to 10 % can be absorbed. In addition, the surface treatment positively affects the temperature, relative air humidity, and radiation intensity.

Translucent concrete

Translucent concrete is also called light-transmitting concrete. In order to maintain translucent properties in concrete, specially developed webbing made of slightly conductive glass fibers is added to the concrete. The fibers are between 2 microns and 2 millimeters, and account for approximately 4 % of the material. Since the share of fibers is so small, the strength properties of the concrete do not change. The fibers are not added to the fine concrete mix but rather deposited into the formwork layer by layer. The distance between layers is about 2 to 5 millimeters. The slightly conductive glass fibers allow us to see light, shade or even colors through the concrete. Nowadays, this material is not only used for internal applications but facades as well.

Breathable concrete

The idea of Breathable Concrete or Air Permeable Concrete (APC) was originally invented for the purpose of developing dynamic insulation for buildings. Fresh outdoor air is pushed through the massive layer of APC in the building envelope and is heated or cooled before being introduced into the building. This process results in better indoor air quality (Wong, Glasser et al. 2007).

The concrete mix developed by Imbabi et al. at University of Aberdeen in Scotland and published later as a patent, required special internal concrete structure. First, high porosity (void ratio) of concrete was necessary. Second, the voids had to be interconnected in order to provide a continuous flow of air through the concrete wall. Finally, regular pore size distribution had to be achieved (large amount of small voids increases the turbulent flow compared to smaller amount of larger voids when the flow is laminar) (Kacejko 2011). Current experiments on the material show that APC is 24 times more permeable than traditional concrete. However, there are still limitations on how far the material can be taken as a structural material since permeability leads to unacceptable losses in strength and structural integrity. Modifications on the nano-scale promises a step forward in the material technology as it might allow the internal structure of the material to maintain a high structural strength together with high permeability properties. These modifications can offer a new generation of concrete material that is lighter and stronger with more functionalities and lower embodied energy (Imbabi 2014).

Gradient concrete

With regards to recyclability and CO_2
reduction, research is done at the
"Institut für Leichtbauen Entwerfen
und Konstruieren" (ILEK), "Werkstoffe
im Bauwesen" (IWB), and "Institut für
Systemdynamik" (ISYS) of the University
of Stuttgart around the topic of graduated
concrete. Building parts are developed
that adapt their material properties
exactly to local requirements. Thus, in
areas with higher insulation requirements
more aggregate is added, resulting in
lower density, than in load-bearing areas
with a significantly higher density. The
manufacturing process for such building
parts involves spraying the concrete into
the formwork with a simultaneous
spraying method. The information about
the amount of aggregate to be added is
fed directly from the computer based on
structural calculations.

2.3 THE OPPOSITE OF HEAVY

People often react funny when hearing about a real concrete canoe (see chapter 4 Projects). Learning that an entire regatta is held solely for this type of boat makes them wonder if this is a joke.

True, regular concrete is heavier than water. Only the density of light-weight concrete (900 kg/cbm) is lower than that of water. Concrete can vary from 1,800 kg/cbm (technically this would belong to the group of mortar or floor screed) to 2,700 kg/cbm (such as for Ultra High Performance Concrete), and regular concrete has a density of approximately 2,000 kg/cbm depending on the material composition and the amount of reinforcement. Big container ships are made of steel which is at least three times as heavy as concrete. Density can therefore not be the only relevant parameter.

Steel as thin as only a few millimeters can absorb strong forces. Concrete on the other hand is not able to deal with tensile forces and requires reinforcement. Since this is most commonly done with steel (load-bearing fiber-reinforced concrete is still not allowed in many countries; Germany, for example) rust protection is necessary which requires a non-corroding material or a cover layer to keep humidity form entering and weakening the steel. The decision is driven by efforts measured in price; stainless steel is far more expensive and difficult to process. Using more concrete than actually needed for the structural requirements is the cheapest and easiest way.

Concrete is liquid during the production process and reaches its mechanical strength after drying. Metals can also be described that way, but the effort to reach a liquid condition is significantly higher. Furthermore, formwork (for simple geometries) for concrete is very easy to produce and can be applied onsite.

The change of state of the material enables the planner to form concrete into any desirable shape. Unlike a rigid material (during production) it can be customized to the individual force flow of a building. Therefore, the use of material is very efficient.

This idea is not new. In the 1920ies, Nervi investigated the force flow and generated construction that guides the force by using only the minimally necessary amount of material. Floral patterns evolve and generate an architectural beauty, even today.

These complex constructions were made possible with carefully engineered preparations. Each project required detailed planning of the building element and constructing the formwork itself. Formwork material and its substructure had to be assembled individually for each part. This was only possible due to the low labor costs of the time. Back in these days the reduction of material resulted in reduction of the budget.

The change in the relation of labor and material cost has changed concrete construction.

Today, wages are high and the material is more cost efficient. That results in a cost distribution for a concrete slab of 80 % for the formwork and 20 % for the material itself. The high formwork costs lead to simplified profiles; ripped and coffered slabs are more material efficient but the formwork is too complex in terms of planning and realization.

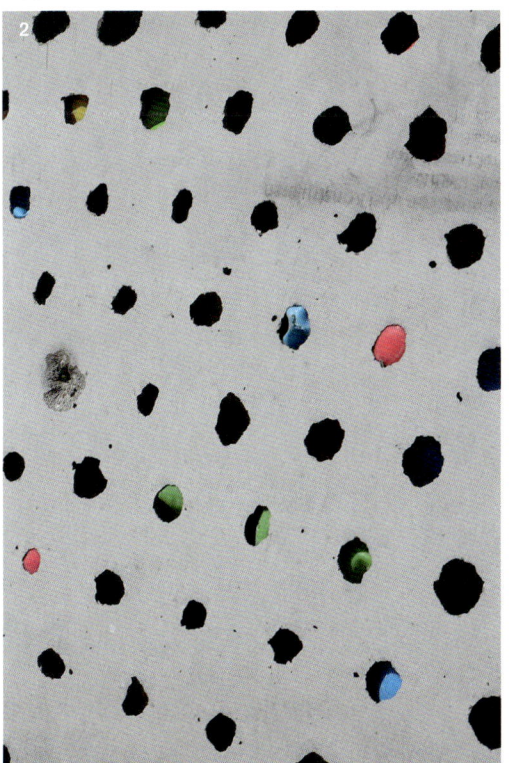

1 Palazzetto dello Sport, Pier-Luigi Nervi
2 Bubble Wall
3 Concrete sculpture by Hickert

4 Copenhagen Business School (CBS)
5 Detail stairs CBS

Light-weight constructions are rather seldom although they have a positive effect not only on the building element but on the entire building context. The load-bearing construction can be reduced with a lighter weight slab construction. A nice example is the Design School Essen. The vertical loads were optimized with the Bubbledesk slab. The structural requirements were reduced and the walls could be more slender.

Another approach to optimize weight and loads was to reduce the slab thickness while improving the materials structural abilities. In the last 60 years significant developments have been made. It is now safe to stand on a balcony with a floor only 70 mm thick due to the impressive performance of UHPC (Ultra Hard Performance Concrete).

The formwork, however, did not change to that extent. It still requires high manual effort. Modular systems and coated wood panels helped optimization but did not essentially change process. Slip form enabled faster construction periods but did not increase simplification.

Looking at the history of concrete in the building context (a detailed view is given in chapter 2.2), effort and respectively money were the main drivers for development.

Today, another parameter is gaining importance. The ecological dimension changes the perception of material. We might be able to supply ourselves with energy but resources are limited. The awareness of the cycle of materials increases and with this appreciation for certain materials grows. Materials should be used were they fulfill a necessity and should not be installed carelessly. Prefabrication makes sense when one element is used and / or made multiple times. Conditions in the factory are customized to the production process and enable qualitative products.

Not only the production process has potential for optimization; waste from mineral material accounts for the highest share of building waste. One approach is to install reused concrete elements in order to keep them from filling landfills and to prevent primary resources from being exploited. The reuse potential of prefabricated elements is higher compared to in-situ concrete building elements as the grid helps the disassemble process. This is only one approach for optimizing the effect on nature. Turning the discussion in the opposite direction – that is how the sustainability debate will impact the material concrete will also become increasingly important. The sustainability thought impacts the perspective and concepts such as "form follows function" are perceived in a new light.

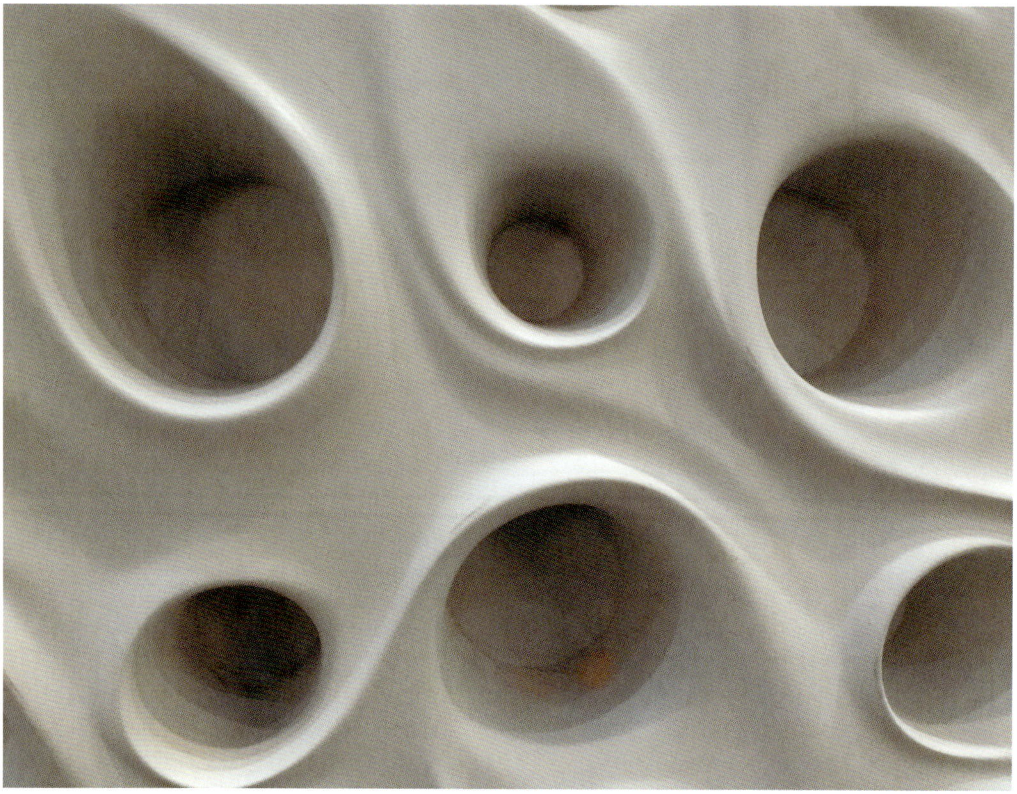

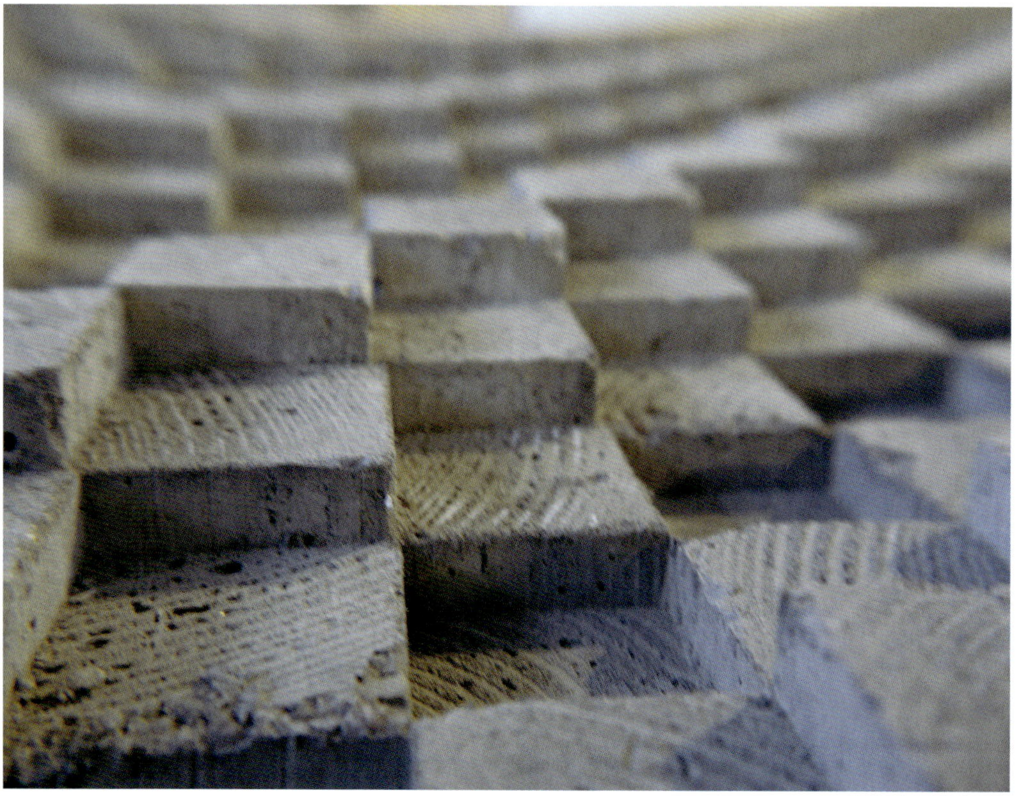

2.4 THE FORMWORK'S MIRROR

Asking non-architects for their opinion on concrete, the answer will most likely not contain very positive attributes. One of the first images that come to mind is that of the parking garages of the 70ies, rough structures that time did not treat very kindly. Or the systemized concrete buildings, the "Plattenbau" which became mass production after the war.

The visible surface was defined by the functional requirements and beyond that did not contain any more information. Meanwhile, architects used the formwork panel and its dimension to rasterize the façade. Furthermore, the formwork anchor became an ornament making reference to the construction process. Beyond this, using concrete's qualities and capacities is an exception or happens on an experimental level. The potential of concrete is far from being exploited.

The potential is the change of aggregate state; concrete is liquid during the building process, and able to adapt to almost any desirable form and even more, to a broad variety of surfaces. A concrete material can not only show different shades; it can appear flat or deep, rough or smooth and even the felt surface temperature can differ.

The desired surface is created by the formwork material, formwork technique, the ingredients, or the post-treatment.

The formwork material has a major impact on the surface. The resulting color shade of the concrete is mostly influenced by the formwork material's absorbency. A surface that absorbs liquid well will generate a darker concrete color. A good example here for is un- or minimally treated timber. The fibers soak up water. A formwork material that repels the water and keeps it at the surface will generate lighter shades of concretes. A laminated wood panel, for example, and even more so, a non-permeable foil or a polycarbonate panel. In addition to the color, the formwork material influenced the haptics. While a rough timber panel shapes the concrete so that the fiber structure can be recognized afterwards, a polycarbonate panel results in a shiny surface, which reflects light. The reflectance creates a fascinating effect and irritates the spectator for a minute, making him wonder whether this is really concrete. Concrete has to flow very easily, thus its ingredients have to be rather small. If this is given, concrete is able to react and adapt to the formwork surface qualities, and will result in a surface that mirrors that of the formwork material. Additionally, the absorbing capacities of the formwork result in less air encapsulation and respectively fewer holes. The concrete is firmer. A non-absorbing formwork surface results in a smooth concrete surface but tends to be more vulnerable to damage.

Another factor influencing the quality of the surface area is the texture of the formwork material. A neoprene fabric, for example, creates a surface area with miniature juts and recesses. Touching concrete that was formed with neoprene is very pleasant as it feels slightly warm. It certainly appears warmer than, for example, concrete shaped with foil. Due to the tiny variations (if you look closely you can see them with the eye) the surface area coming in contact with the skin is smaller compared to that of the foil example. Air is stored in the recesses, and what our hand feels is partly concrete and partly air, which feels warmer than just concrete.

Formwork material that results in light and dark areas can be used next to each other to create a type of typography on a flat panel. Since concrete reacts differently to different surfaces, there is a broad range of design possibilities. Beyond this visual aspect, concrete can incorporate different depths and include a third dimension. The Pixel Façade shows a three dimensional concrete façade board. 30 x 30 mm timber blocks (massive timber, sawn and sanded), 10 – 50 mm high, were arranged in certain pattern. In order to take the wood blocks out of the concrete the maximum height difference from one block to another was limited to 10 mm.

In initial experiments the blocks were bathed in mould oil in order to repel the water. The texture of the wood was very visible and a little bit sandy. However, the mould oil helped only little as humidity barrier. Taking out the framework was still difficult. Spraying the timber blocks with varnish helped to loosen the connection between formwork and concrete. Another experiment showed the potential of a silicon layer in between positive and negative forms. Disassembly was rather easy, and the silicon matrix could be reused several times. The concrete does not have the timber texture but is very smooth instead and mirrors the silicone layer. The color varies from one example to the other. The concrete shaped with timber formwork is darker compared to that shaped with silicone even though the same concrete ingredients were used.

The desired visual quality is the first to be defined. The ugliness of concrete facades is not necessary; streaks occur with time due to the exposure to weather conditions and a lack of maintenance. A rough surface supports slow water flow and leaves unfortunate spots.

Architects never lack challenging ideas when it comes to facade surfaces. We begin to understand that the surface and concrete ingredients are reflected on the concrete's surface. These are relevant parameters for the construction phase. The work continues after that; we need to understand that a concrete façade requires the same effort as a plastered exterior. It needs to maintenance. Let's design surfaces we love so we take good care of them!

Formlining – different concrete surface results

3. FUTURE PERFORMANCES

3.1 INTRODUCTION

The topics construction (own weight, thermal expansion, horizontal stress), construction method (process, production, construction), climate control (temperature, humidity, acoustic, light and view), sustainability (embodied energy, storage capacity/usage energy, recycling, consumption of resources), usability (safety, fitness for use, climate, maintenance) as well as appearance (urban context, exterior and interior surface perception) can be identified as individual functions of a building construction. The purpose of this chapter is to sketch potential areas of future development – to improve the functionality of the construction.

According to active research within the Façade Research Group (see chapter 1) the thermal / building physical performances of the building envelope are an important aspect of investigation. Hereby the main question is which service components to select for a massive building envelope and how to integrate them in order to improve the overall function of the façade and the building.

In addition, the question of how to solve the principle problem of free-formed constructions is elaborated. Related to the research within the group, technologies are sketched that will provide reusable free-form molding solutions. And finally, the chapter focuses on the topic of more efficient molding systems compared to framework constructions. Here the concept is to either identify maximally effective molding principles – which typically means stress loaded systems – or, alternatively, to use the technology of additive manufacturing to build the construction.

3.2 INTEGRAL SOLID FACADES

FROM MONOLITHIC TO MULTI-LAYERED TO INTEGRAL

"Solid facades" is the umbrella term for monolithic, multi-layered and integral wall construction. Most commonly, solid facades involve mineral materials but technically it could include solid wood construction as well. However, since this is a concrete book we will focus on mineral material.

The wall started as a monolithic construction, and when additional functions were required it became multi-layered. One approach for an ideal construction is to combine the two; the simplicity of a monolithic wall with the functionality of a multi-layered one – the integral solid facade.

Monolithic facades

A monolithic wall consists of only one layer and is made in one procedure. This offers many advantages; the reduced material variety simplifies the production process and the installation of the building element onsite. The risk of failures is reduced and the planning as well as the erection process works very efficiently.

Quarry stones and bricks are traditional massive materials for wall erection. They possess many important features for exterior walls such as strength and durability. In the 1920ies, a house-builder in England started to include an air gap in between two solid layers to overcome the water penetration problem. This was adopted by builders from that time onward (Burnett 1986, Scaysbrook 2011). This type of construction had an additional positive effect as it contributed to vapor tightness. The moisture was kept from penetrating the inner solid skin and helped the moisture dissipate back to the exterior. Whether a wall of two solid layers is regarded monolithic or multi-layered is a point of discussion. The definition we offer differentiates the two by the process steps involved to construct them. We would call it monolithic if the two layers are made within one process and multi-layered if one is built after the other.

When walls did not have a certain thickness, they allowed water and vapor to penetrate which initiated the use of render. Technically, this was the first step toward the multi-layered wall. Increasingly higher standards for the interior space required higher functionality of the exterior walls. This initiated the layering of functions.

The monolithic facade served mainly as a load-bearing layer as well as thermal insulation, water barrier and fireproof layer according to former necessities. The standard of the functional require-ments kept increasing essentially over time, and today a monolithic wall is no longer a common construction type due to high physical requirements and its functional limitations.

Multi-layered facade

The development from monolithic to multi-layered can be traced on a small scale (like the addition of render) or on a more visible string of action.

In the medieval era churches were built with great height and slender walls to let more natural light enter the interior. The

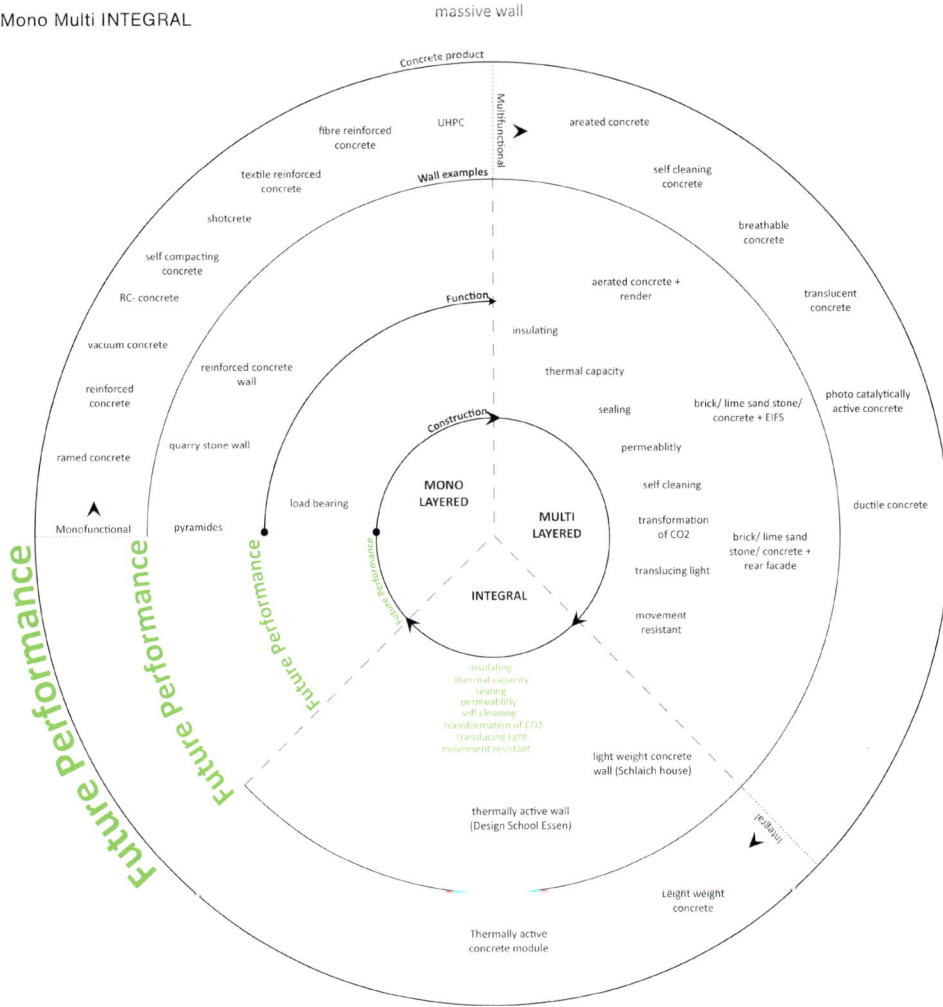

Mono Multi INTEGRAL

flying buttresses system was developed to split the functions; carrying structural loads and separating inside and outside were served by two different building elements. This can be perceived as one of the first preparation for the separation of functions and as the first multi-layered wall. The separation of functions became even more radical with the development of steel and its wide application in the building industry in the beginning of the 20th century.

The oil crisis in the 1970ies fundamentally changed the thermal requirements of facades. A lighter material was added since solid materials such as brick, sandstone or concrete have poor thermal resistance. This insulation layer needed to be protected from weather conditions which made an additional layer necessary.

Integral wall

In the old days, when standard requirements were lower than today, the designer solved the functional challenge with only one type of wall. That kept the planning process simple and he/she had to supervise only one craft; for example, a quarry stone wall solved all functions in one. We do not want to go back to those living standards but the idea of solving all problems in one process step is very attractive. What we want today and what we expect to be possible in the future is a wall system that offers a high variety of functions and can, at the same time, be produced in with a single simple and efficient production step – the integral wall.

A wall that is perceived as monolithic would offer many advantages. Neither the architect nor the building owner has to choose a material mix or type of connection if there is an all-in one product. The reduction of information (less materials, less joints,...) offers a design quality. The failure rate could be reduced as the number of crafts onsite is minimized. Additionally, the planning process could be simplified and surplus capacities can benefit the overall design.

Today, we can already witness some projects that demonstrate the potential of the integral wall. They focus on the thermal performance which can be distinguished by active or a passive approach. Both are briefly introduced with one example.

Passive thermal performance – Mike Schlaich House (2007)

Again it is the monolithic appearance of the Mike Schlaich House in Berlin that gives the first impression. The outer skin meets the thermal requirements without an additional layer. This is possible due to the concrete product Infra-Lightweight Concrete (ILWC, developed by the Technical University Berlin). As the name implies the concrete is lighter than regular concrete due to extended clay additives. At 50 cm, the facade is comparably thick but has no recycling issue at the end of the life cycle as a common EIFS would have.

Active thermal performance – Zollverein School (2006)

The active approach uses local resources in order to meet the functional requirements. As thermal insulation is a major issue, harvesting renewable energy eliminates the need for an additional (passive) layer. This helps to decrease waste during the end of the usage phase and minimizes the material and joint variety. This concept is especially sensible when renewable energy is available within short distance.

The outward appearance of the Zollverein School is very straight. The observer recognizes only one material in a slender construction making it appear very elegant.

The design of the Japanese architect SANAA included very slender wall constructions. In order to build these in Germany (and fulfill thermal requirements) insulation was needed, but common sandwich constructions would increase the cross section significantly. An alternative was found by using an active insulation. Fortunately, the project was located near a former coal-mine, and free hot water from the mine was continuously pumped into the façade.

The integral system provided a 30 cm monolithic solution, almost half the thickness of the regular layered façade. Additionally, it optimized the use of site resources by harvesting hot water from the coal mine. Finally, it enhanced indoor quality. Thermally activated surfaces provided better indoor quality compared to conventional forced-air distribution system.

1 Integral concrete column with a wound
2 Zollverein School of Management and Design, Essen /
  SANAA architects
3 Jakobitor, Soest
4 Ultra-lightweight house, Berlin (Credit: Maik Schlaich)
5 Phæno – Science Center, Wolfsburg, Zaha Hadid
  architects

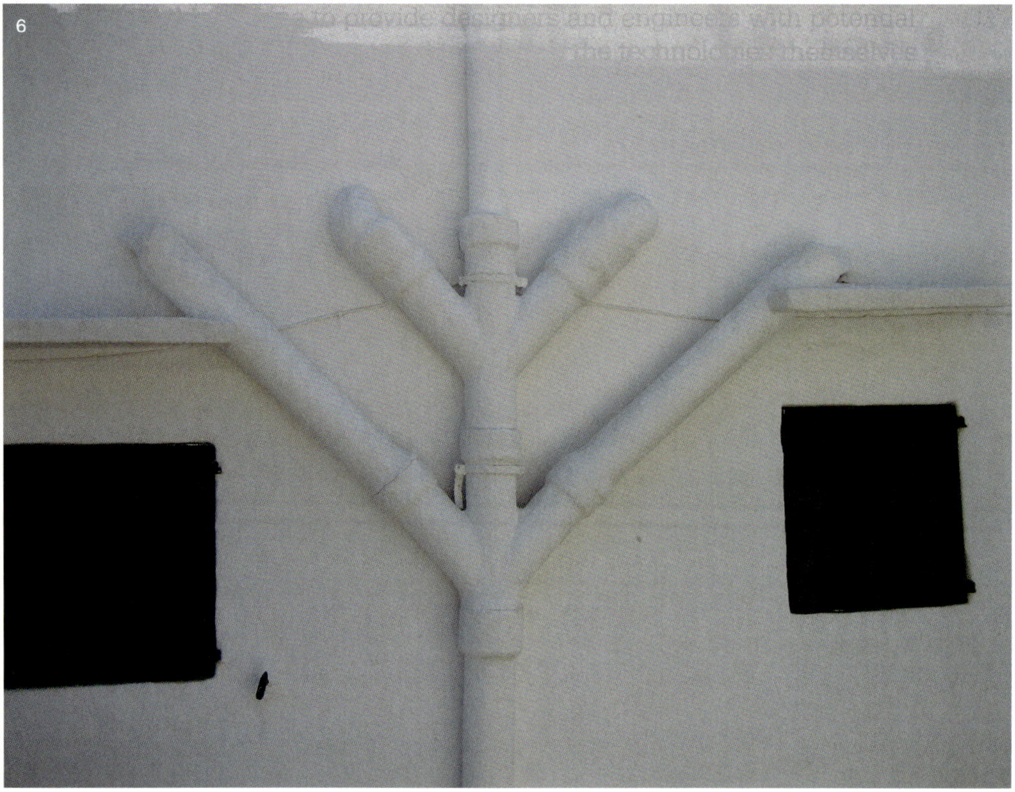

6 Downspout, Menorca
7 Crematorium, Berlin, Schultes Frank architects

What next?

Solid wall constructions today integrate more functions while still appearing monolithic. On the material level, new concrete technologies are being explored; promising new functions and solutions for massive walls are changing the way the material is used. New technologies have been introduced for concrete walls. Today they are able to bear all kinds of functions.

The monolithic appearance of concrete walls offers an aesthetic quality. Maintaining this quality and exploiting its functionality needs to be pursued for future applications. Integrating more functions into the massive component is likely to happen by various measures, as shown in the previous examples. It can be by changing the material composition, and in this case the material itself gains an extra function. Another form is by adding extra components or systems to the massive layer. In this case the components work together in order to allow the addition of functions. Another form is changing the shape/form of the massive layer to gain an extra function. The common perforated blocks are an example for this type of measure. By adding voids in the massive block the layer gains insulating properties without changing the material composition or adding extra components.

The monolithic wall bears a strong architectural attraction as it relates to the original idea of a wall. It does not communicate any functional information. It can be compared to a MacBook with its complex machine and clean aluminum shell. Our approach to design is of an unobtrusive nature. The functionality is not shown which is regarded as elegant.

This simplicity has a second meaning; form the architect's perspective less communication is needed. Instead of coordinating various companies he / she only has to speak to one. Having less partners decreases planning and communication time.

We want both: high functionality in a facade and simple systems. High performance facades are mostly realized by multi-layered facades (also component facades). Not all functions are needed in all surface areas but the facade needs to work as one coherent and smart system.

Further steps will include integrating not all but a limited amount functions to a monolithic appearing system. Maybe the integral wall will start with the application of complex components layered in the facade, like a multi-layered facade covered with a monolithic shell. Based hereupon components can be reduced and the production process can be simplified. The ideal is to have a monolithic (or integral – you are welcome to add your own definition of the two) wall that serves all desired functions.

3.3 ADAPTABLE FORMWORK FOR FREE-FORM WALLS AND SHELLS

Concrete is the structural free-form material per se. In fact there is little other material suitable for structurally covering and enclosing space at the same time, while leaving ample room for architectural expression. When brought into an appropriate form, concrete shells feature high performance and unique aesthetics. However, we find metals and wood predominating contemporary projects as these materials do not require formwork. Examples of avoiding concrete when it comes to free-form geometries are as old as the material itself.

In 1917 Mendelsohn's expressionistic design of the Einstein Tower was inspired by the possibilities of the formal freedom of reinforced concrete. The building was later erected using masonry because there were far more qualified bricklayers than carpenters available to realize a wooden formwork. The era of the concrete shell realizations by Oskar Niemeyer, Felix Candela and others in South America seized abruptly when wages of union workers where drastically raised in the nineteen sixties, making it too expensive to fabricate specially shaped formwork for in situ casting.

The designs of digital liberation brought back free-form concrete architectural elements. Nowadays, formwork is manufactured using computer aided technologies resulting in rib and bulkhead style wooden formwork as it was used

for Sanaa's EPFL Learning Center or milled foam blocks assembled to a formwork as they were used for various projects of Frank Gehry in Dusseldorf, Germany. Like their predecessors of the shell area these negative molds are one of a kind installations; meant to be used once and to be demolishes after use, ignoring aspects of resource effectiveness. The enormous material and labor effort actually inhibits the realizations of architectural concrete elements such as long shallow curved walls. In an analysis by Hickert[1] it was outlined, that this type of geometry was hardly in existence at all, presumably due to the cost for a large scale tailor-made single use formwork.

A formwork that is adaptable to various geometries and that can be used multiple times will have a strong impact on the availability and cost of free-form concrete walls and shells. Especially for concrete walls there will be an important impact as they have the highest production cost of all concrete parts.

Presently, various patents exist and we find early implementations of adaptable formwork tables. However, little research is done in this respect and the ongoing research focuses on horizontal table-like systems. None of the present developments have been realized to a stage where structural concrete elements could be cast. We find a promising realization approach at the Danish startup company Adapa SpA, where adjustable formwork tables are being used to cast foam blocks that are to be used as formwork components, eliminating the milling process but still leaving room for improvement towards resource efficiency.[2] Research is presently conducted by Shippers at TU Delft, focusing on horizontal systems.[3]

1 Einstein Tower, Mendelsohn, 1922
2 Construction site of the National Congress, Brasilia
   (credit M. Gautherot)
3 Hard to find: Double curved in situ cast walls
   – Ceramics museum, Laufen, Germany
   (credit Nissen & Wentzlaff)
4 Esthetics of shallow double curvature: Gina Design
   (credit BMW Group)
5 Formwork table by Adapa ApS (credit Adapa)
6 Research on concrete technology for adaptable
   formwork (credit Roel Schipper)
7 Spline ruler with spline weights (credit Carl de Boor)

In our approach, research is focused on adaptable and reusable formwork for in situ casting of free-form concrete walls. Besides cost reasons, we focus on walls because they are more relevant to contemporary architecture than horizontal parts. The in situ casting is an important step towards making full structural use of shell action that free-formed walls usually offer. Finally, the pressure of the concrete being poured into an upright formwork is by far higher than that of a horizontal one, so research results of formwork surfaces capable of withstanding high concrete loads can be transferred to horizontal formwork tables any time. The system architecture of an upright arrangement is different from formwork tables, opening up a new field of research.

The adjustable formwork for walls will have at least on set of hydraulic or mechanically driven actuators to imprint a specified geometry into the formwork surface. This process is electronically preprocessed and controlled. The form-work surface stands upright inside a supporting frame and will be held in place by the actuators that are firmly attached to the surface material. Different layouts of the formwork surface arrangement can be imagined. To cope with the high concrete pressure and still preserve the ability to deform the surfaces elastically, a system with two anchored surfaces is most suitable to solve the problem as the anchors shortcut concrete pressure and slave the exterior mould surface to the inside that is driven by actuators.

It is important to understand the principles of elastic deformation of flat materials. Free-formed surfaces can be understood as a dense sequence of spline curves, representing the surface. Today, the spline is firmly associated with the CAD technology but it originates in the ship building culture when the word described a thin and flexible ruler, which was elastically deformed to interpolate between defined points resulting in a smooth curvature. The spline was held in place by spline weights, acting as control points. The ruler's path can by described by a spline interpolation of minimal bending energy and amount of curvature. In relation to our formwork surface a spline ruler can be understood as a fiber of the elastic mould surface material and the physical spline weights resemble the control points of the spline, represented by the actuator tips imposing deformation on the surface. While splines in computer programs may encounter an arbitrary sequence of curvatures, the materialized spline is limited by its mechanical

Various Working Principles of adaptable wall formworks

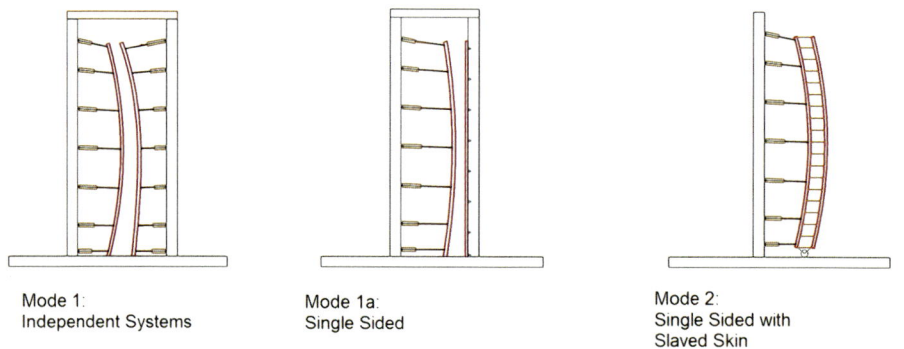

Mode 1:
Independent Systems

Mode 1a:
Single Sided

Mode 2:
Single Sided with
Slaved Skin

properties. For approximation of curve or non-planar surface by bending, control point spacing, stiffness and external loads are of influence.

For the configuration of an adaptable formwork surface there are four vital interacting parameters that have to be considered:

- Actuator distance
- Given target curvature
- Surface material stiffness
- Concrete pressure

The above points influence the configuration of a mould system and its ability to approximate a given topology. It is one of the central tasks to define configuration rules and performance limits for the dimensions of above parameters.

The behavior and interaction of actuators and formwork surface proved to be of higher complexity than the simple theory would make it seem: Simple formwork tables such as the Adapa type feature strictly vertical moving actuators. Obviously the surface material cannot be firmly attached on the actuators because once it is deformed to a curvature; it will slide on the actuator tips to maintain its original length.

If the actuators get attached to the surface, they will have to follow the surface's movement and will incline. A horizontal formwork table can work either way, sliding or attached. But for the upright system, attachment is mandatory. The actuators stabilize the upright surface and since being attached to it, the actuator tips have to follow and incline against their initial position, resulting in an inclination.

What's so tricky about inclining actuators? The physical implementation is simple, once the formwork surface is kept from moving sideward. But to maintain accuracy, the processing gets a bit complicated. To investigate the interaction of tilting actuator and formwork surface, a simulation environment was created that anticipates all movement of the actuators, simulates the presence of the formwork material, calculates various compensations and finally features a data interface to control a functional mock-up of the formwork.

Here is just a sketchy description of what happens during the control of the formwork. If we want the actuators to imprint a double-curved spherical form into the formwork surface, the following steps will be performed: A grid is projected on the target surface and the

Patent sketch by Adapa ApS (Grahic: Patent US 20130299084 A1)

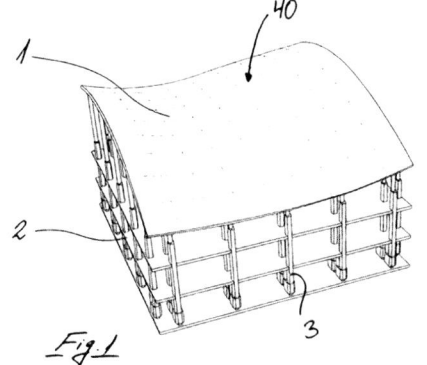

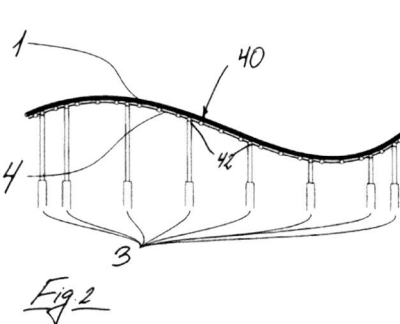

initial actuator stroke is calculated. The actuator movement is then simulated, resulting in inclined actuator axes and a curved surface representation. Now that the outer actuators have inclined, the curvature of the surface deviates from the intended geometry. The actuators' stroke has to be compensated so that the surface gets approximated back to the wanted geometry. This process is adjusted repeatedly until convergence is reached. Finally, the compensated actuator stroke is communicated to the USB port and handed to the actuators that are listening for individually addressed commands they respond to.

It is the nature of any flat material to resist deformation into a double-curved geometry. The effects taking place are easily visualized when trying to drape a sheet of paper over a spherical object, e.g. a ball, or one's knee. The edges tend to wrinkle under compression while the center is likely to tear because of the tensile stress. If a woven textile is used instead, the corner areas show that the fibers suffer a rhombic distortion. So we can say, a flat material deformed around two axes encounters planar compression and/or tension as well as shear.

A perfect surface material would be one that can be easily stretched, compressed and distorted. On the other hand, we want the material to be as rigid as possible against bending action to withstand the concrete pressure. At first glance, axial elasticity combined with high bending stiffness seems to be a contradiction. Like always in such situations a good compromise has to be found.

For mold making, special hole patterns are known to make sheet metal drapeable around double curved objects. A product known under the brand Formetal® inspired to investigate the potential of this technique. It features a structure of Y-shaped holes leaving triangles with cambered interconnections. Under tensile force the triangles rotate around themselves; opening the Y-shaped holes and allowing the material to stretch. Based on this experience a series of patterns with similar behavior was evaluated.

Parameterized patterns are created using Grasshopper for Rhino, featuring square cores and four chambered interconnections. Under axial force the cores will rotate in bent interconnections, allowing elastic deformation. The square layout allows distortion under shear. After the

The usable format for the studies shall be 8 x 3 m, the radii are limited to 8 m and 16 m respectivly

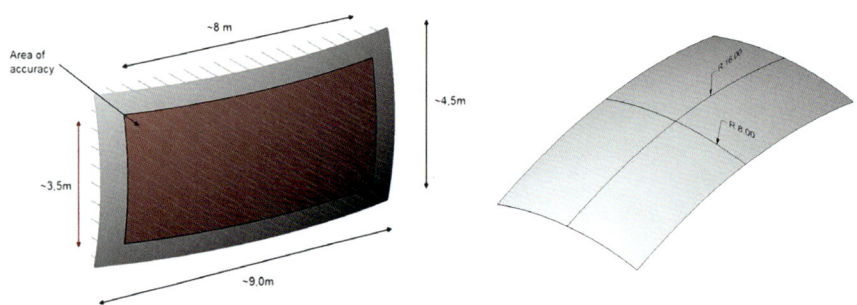

pattern is parameterized, its structural performance can be simulated by adding routines that create a structural model and feed it into a numerical solver. With this arrangement any the structural performance of any parameter change will be fed back instantaneously.

By changing single parameters increment-ally, performance diagrams are created, helping to identify an optimum pattern to preserve a decent bending stiffness in relation to the best possible deformability. The pattern with the rotating cores features at a certain configuration a 90 % reduction in deformation stiffness while bending stiffness is reduced by only 60 %.

Once a pattern is found and optimized it is time to verify if a formwork surface can be created using the results. For the above example pattern a true scale structural analysis was performed, using the material properties of a soft polymer 10 cm wide with steel cap-strip inlays to improve bending stiffness. For an actuator spacing of 50 cm the formwork surface was proved to withstand a concrete pressure of 3 tons per square meter which is quite a realistic value for wall formworks.

CONCLUSION

Nowadays, free-form CNC milled or router cut formwork makes extensive use of material resources. A reusable formwork with the ability to change its topology multiple times according to a given geometry has the potential to significantly reduce the economic and ecologic impact of free-formed concrete walls. It makes expensive single use formwork obsolete and will allow for an extended formal freedom in architectural design.

The created simulation platform can couple geometric and structural analysis, thus making the behavior of complex actuator-formwork surface predictable. Insights were created about functional principles as well as performance data extracted that lead to configuration methods for the physical implementation of an adaptive formwork. The evaluation of customized perforated surfaces delivers promising results and leads to first feasible dimensions and material selections. A physical mock-up is under construction and will be used to perform calibrations of the simulation model.

As the formwork material's length is unchanged, it will make the tilting actuators follow;
Right: Inclining actuators result in a change of the curvature: compensation of the stroke is required

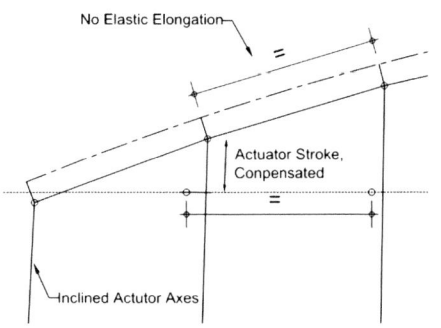
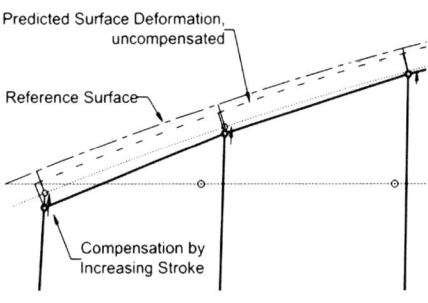

8

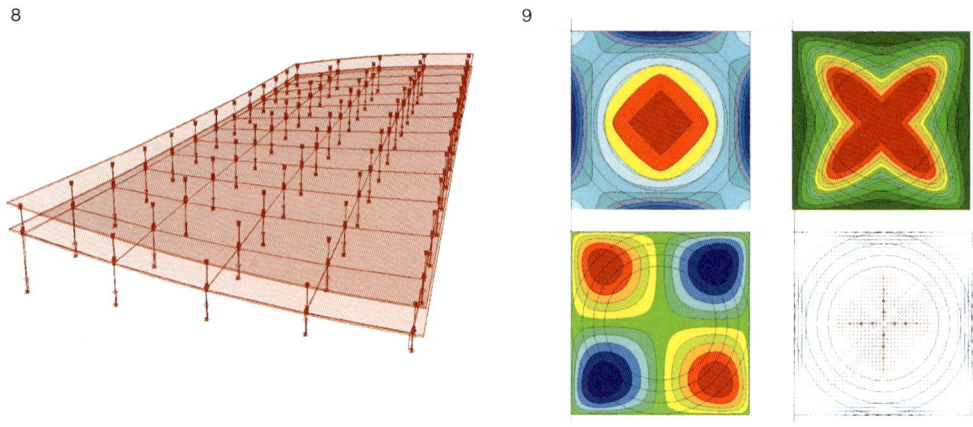

9

10

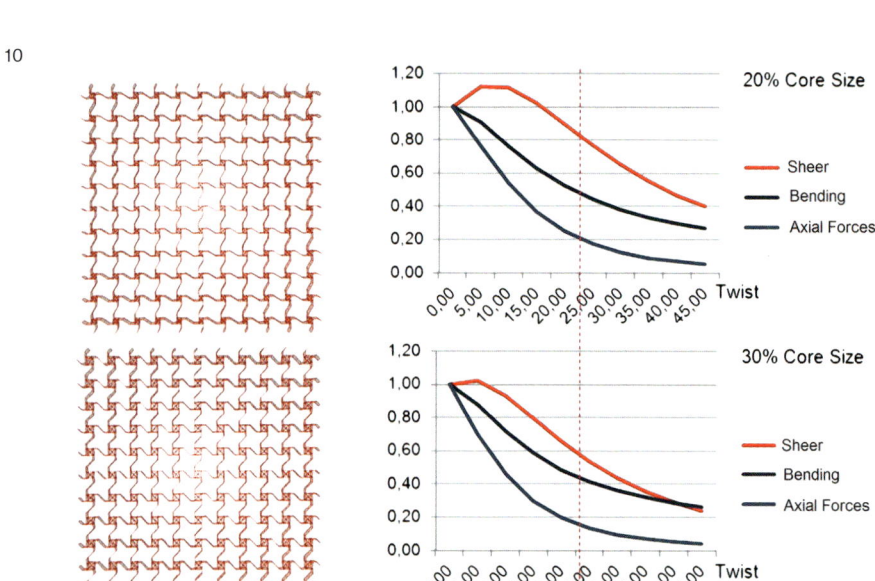

REFERENCES

1   Hickert, S. (2013). Evaluation von Fertigungsverfahren
    für Freiform-Beton-Sonderschalungen in Bezug auf
    deren Weiterentwicklungspotenziale. (Master Thesis)
    Hochschule Ostwestfalen Lippe, Faculty of
    Architecture

2   Raun, Ch., Kristensen, M., Kirkegaard, P.H. (2011).
    Dynamic Double Curved Mould System. In: Gengnagel
    et al. (Ed.) Computational Design Modelling,
    Proceedings of Design Modelling Symposium, Berlin

3   Schipper, R., Grünewald, S., Raghunath P., (2013).
    Rheological parameters used for deliberate
    deformation on a flexible mould after casting,
    Proceedings of RILEM 1st Conference on Rheology of
    Construction Materials, Paris

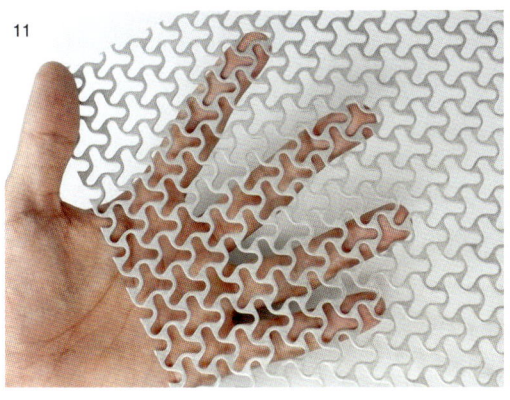

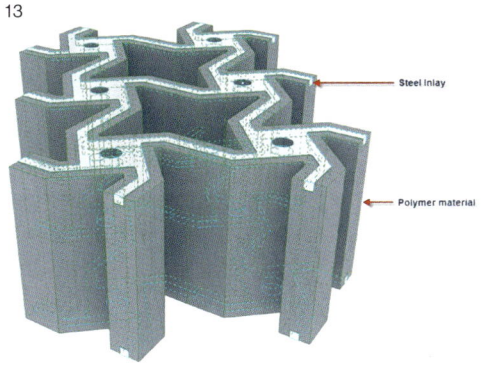

Steel Inlay

Polymer material

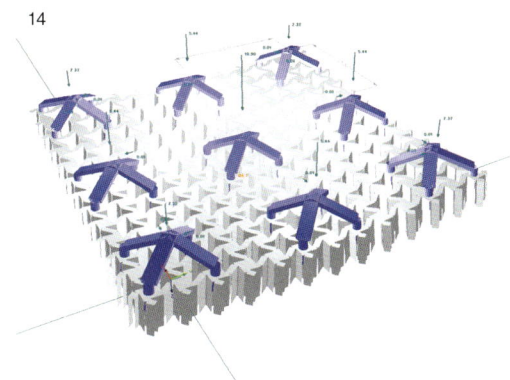

8 Simulation model of two layer formwork
9 Stress analysis of deformed formwork surface
10 Analysis of the pattern under various rotation
   values: Axial stiffness decreases more rapidly
   than bending stiffness
11, 12 Deformable hole pattern of Formetal®
   (photos: DINOSAURIER-Werkzeuge Trading GmbH)
13 True scale visualizations of the analyzed
   formwork surface
14 Proof of concept: Structural analysis of true scale
   implementation featuring 50 cm actuator spacing
   and 10 cm perforated soft polymer formwork
   surface material

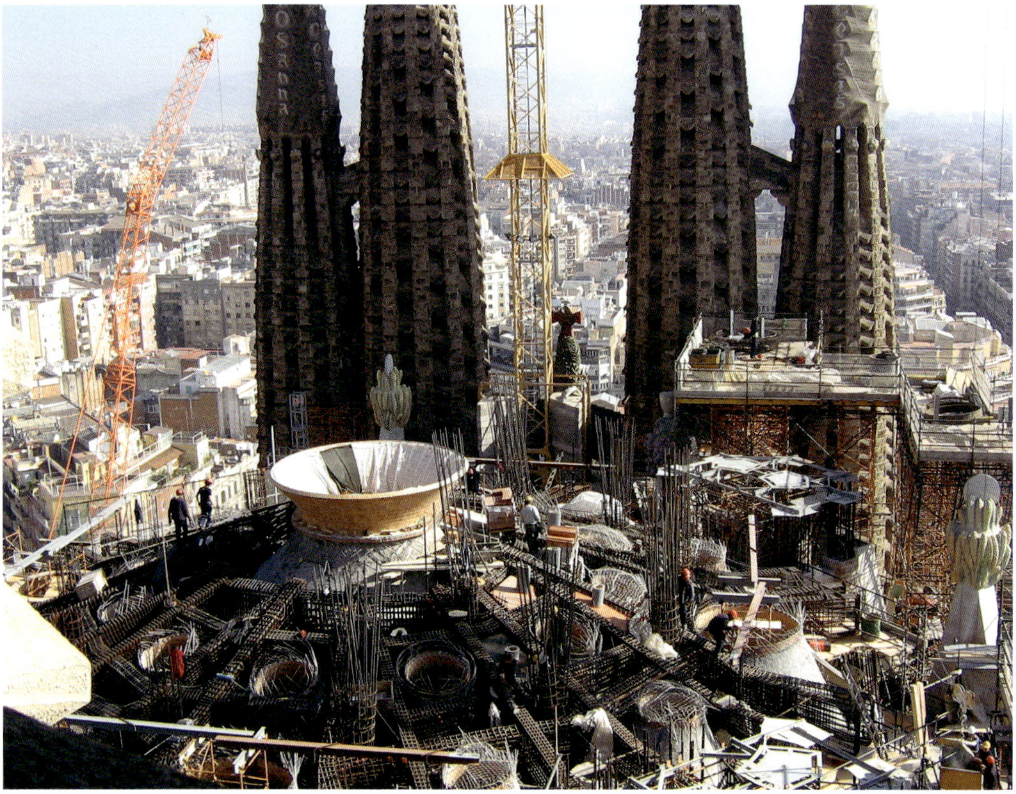

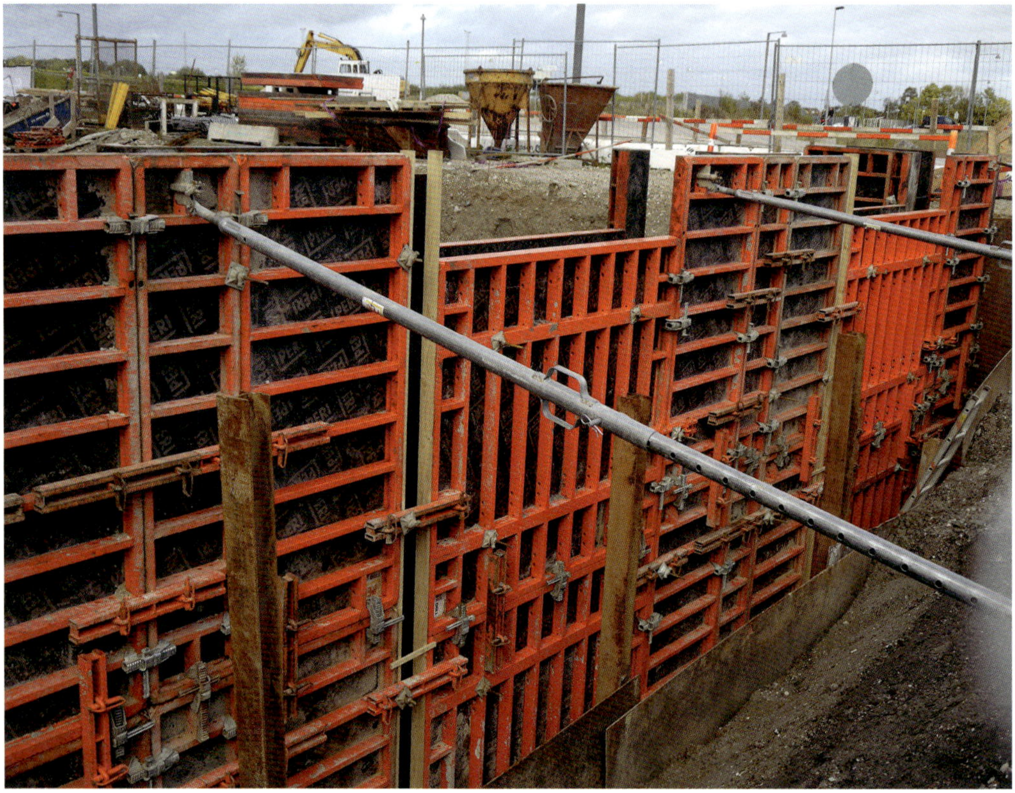

3.4 BEYOND FORMWORK

When examining the development of conventional steel-reinforced concrete it becomes noticeable that it took until after World War II for a significant and lasting development and rationalization phase to occur. Initially, rationalization was limited to the topic "reinforcement and concrete". First successes in working more cost efficiently were the result of using bending facilities for the reinforcement and the development of machines for concrete production with high mixing and pumping rates. However, these rationalization measures were soon exhausted and the focus shifted toward the formwork, a hitherto much neglected part of the process. Initial improvements and approaches referred to pre-planning, new operating resources (prefabricated formwork elements) and possible areas of application (Kowalski, 1977).

Additional developments:
- loose boards were successively replaced by formwork panels
- formwork panels were replaced by system formwork
- wire was replaced by tie-rods
- screw connections were replaced by rod strainers
- round timber columns were replaced by length-adjustable steel pipe columns
- squared timber was replaced by glued laminated timber, steel or aluminum girders
- steel girders were replaced by length-adjustable steel girders
- "formwork construction" was replaced by "formwork technology".

In spite of all developmental measures, formwork is still dominated by manual labor. Numerous studies (Müller (1972), Rathfelder (1992), Hoffmann (1993), Reichle (2002)) show that even today the cost for formwork accounts for a significant share of the cost of the overall building shell construction for a conventional steel-reinforced concrete high-rise (30 – 40 %). Even with the most rationalized formwork systems currently available.

NON-LINEAR BUILDING DESIGNS

Considering the ever increasing architectural demand for skew, free-form building geometries and the need to produce curved concrete building parts, the above described rigid formwork systems do not suffice.

Established formwork systems to produce curved concrete building parts exhibit a stiff deformation behavior. In as far as the surface curvature cannot be linearized in segments two methods to create curved surfaces are being used. Firstly, general areas are reduced into developable areas on a segment by segment basis, i.e. the formwork is produced with a number of customized wooden boards (wooden console).The smaller the boards, the better the desired shape can be created. Secondly, the formwork for double-curved surfaces can be generated by milling negative shapes (for example in foam blocks), which are combined into a formwork segment with the aid of individual support structures. Inherent to both variants is the fact that they require a stiff form which can be used only once and cannot be modified. One method of creating a fair-faced concrete surface of increased requirements is to laminate the formwork surface with epoxy resin. Pneumatic formwork systems hold an exceptional position. In general, pneumatic formwork means membrane structures that are held in shape by pressure, serving as formwork structure and formwork surface at the same time. The principle is analogue to that of an air-inflated structure. Typically, the materials of choice for such membranes are PVC-coated polyester or polyamide foils. The advantages of this type of construction lie in the reusability of the formwork material, and a fast assembly and disassembly process.

Disadvantageous is the limited application spectrum; this principle only allows to create synclastic load-bearing formwork structures.

EXPERIMENTAL DEVELOPMENT

The most divers protagonists from research, education, engineering firms and the industry have spent many years to identify simpler methods to create formwork for curved or free-formed building parts. These methods can be differentiated into experimental and application-oriented methods. Application-oriented methods are methods that have already been partially implemented in building-unrelated manufacturing methods, and show potential to be used for formwork for the building industry.

At ETH Zurich, students under guidance of Gramazio & Kohler have analyzed the properties and formation processes of sand. The results were to provide conclusions for waste-free concrete formwork. In a study, a robot filled sand into a box, either all at once or gradually, and spread, layered or compacted it. For the method of filling the sand gradually, an industry robot dispensed the sand into a storage container along a predefined path. And at also predefined areas, the flow-through rate of the sand was controlled by means of a control valve.

In contrast to the method where the sand is already in the box; hereby, the loose sand is displaced or shaped with a rotational movement with the help of a tool (similar to a wooden spoon).

The very absorbing surface of the sand poses a problem, as does the process of pouring fresh concrete onto the sand because the sand is being deformed. To prevent this from happening, the sand was fixated with varnish or a latex skin, which also reduced absorption. The method of compacting the sand provides different results. Here, molding sand was used instead of loose sand. Molding sand is an oil-bound or clay-bound type of sand, originally designed for metal foundry. When compacted, the sand is dimensionally stable. Simple pounding tools (wooden plungers) are used to mechanically compact the molding sand. However, the traces of the tools remain visible. Molding sand features an oily surface and in its compacted state, the absorption characteristic can be described as "absorbing". Alternatively, in another project, the molding sand was replaced with clay. The company EEW-PROTEC in Kiel, Germany also examines the material sand. They compacted a soluble glass and sand mix into a block, and then milled a previously generated (CAD) form piece from the block in a computer controlled processing center with different milling heads. This already established method allows integrating tongue and groove connections so that several pieces can be joined accurately. The method, originally developed to produce metal cast objects, offers potential to create concrete parts in a facility for prefabricated concrete parts. The advantage of the technique is the reusability of the soluble glass/sand mix.

The Department of Structures and Structural Design and the IBF (Institute of Metal Forming), both at RWTH Aachen, Germany have developed a transformation method named "Robot-based incremental sheet metal transformation", with which sheet metal can be brought into the desired shape with the help of two robotic arms. This method is not yet used in the formwork industry, but if offers potential. One robot deforms the sheet metal from the upper side with a very small semi-spherical tool, and the other provides the required counter pressure from below. Due to the fact that the piece of sheet metal is clamped in a frame it retains its original outer form, and the new shape is pressed and stretched out of the thickness of the material. This method is derived from the handcraft of peening, changing the shape of a metal object by hammering. This method also promises great potential with regards to creating repetitive formwork elements.

If a positive form exists that is to be cast into the concrete, vacuum thermoforming proves to be a suitable method. During project work at Detmold School of Architecture and Interior Architecture, students Carina Kisker and Maximilian Ernst developed a façade element with the impression of a person freeing itself from that façade element.

With thermoforming, vacuum forming or deep drawing, mainly thermoplastic resin but also glass is shaped into a new and three-dimensional form. The material must be provided as foil on rolls or as sheet material.

The basic principle involves clamping, heating and then reshaping the material by vacuum. After reshaping the generated form is cooled down and post-processed if applicable.

Two different materials were used for this project. As part of the first approach, latex was cold-formed. A gypsum negative shape was generated under vacuum, which served as formwork for the concrete. The advantage of this method is that the material is not stressed to its elastic tensile limit and is thus reusable. The second approach involved warm-forming PE foil. First, PE foils is heated to just under the melting temperature; then, vacuum is applied. After cooling down, the PE foil can be used as formwork material; therefore eliminating the need to create a negative shape. The PE formwork was only supported in a sand bed during concreting in order to prevent possible deformation.

Alternatively, a negative form can be created with the established second cast technique. With this method, a casting mold is made from an already existing model. This is done either with poly-urethane elastomers (PUR) or with silicone elastomers (SI). The particularly elastic properties of these materials allow that one casting mold can be used for a multitude of replicas.

The lamp series "Shadyshades" by Michael Haas and Tom Pawlofsky served the Institute for Architecture and Planning at the University of Liechtenstein as an inspiration to develop free-formed formwork from honeycomb structures. The method involves cutting individual layers of cardboard by means of a computer controlled tool plotter and gluing them together. Due to its low price, firmness, easy and fast processing capability and simple recyclability the choice of material fell on corrugated cardboard.

The biggest problem of the construction was to close off the honeycombs, i.e. the lids.

The final version of the formwork included serialized, tab-like shuttles that were put on the honeycombs; the vertical and dense arrangement significantly increased stability. For easier assembly, the shuttles were connected row-wise. In order to prevent impressions of the honeycomb structure on the surface of the cast piece, the formwork was plastered with a 2 mm thick gypsum layer. The closing of the honeycombs and the surface treatment made it possible to actually walk on the formwork. Finally, a wooden framework was built for the formwork und the corrugated cardboard was protected from humidity with a PE foil.

The result is a formwork product which was initially built as a 2.5 x 4.5 m prototype, and then found standard application in the field of skate park construction.

Another project at the Detmold School of Architecture and Interior Architecture dealt with the topic of elastomer form-work. The challenge that the students faced was to develop a formwork concept, for which the formwork had to be reusable and suitable for changing free forms; that is to say it was to be easily adjustable and guarantee a consistent thickness of the product.

On one hand this type of formwork is interesting due to its flexibility allowing for many variations in form, but on the other it holds several problems. The own weight and the pressure of the fresh concrete cause stronger deformation at the bottom side of the formwork than in the upper areas. There is no undoing this basic physical problem. However, measures to counteract it were tried out: a support medium that absorbs the pressure of the fresh concrete and thus counteracts the deformation in the lower area. The idea was to design a formwork system consisting of a flexible skin and a pressure equalization chamber. Two of these systems placed in front of each other result in a complete formwork system for a free-form wall element. The working principle of such a 3-chamber system allows producing a free-formed wall element of almost homogenous thickness. While filling concrete into the center chamber, the two outer equalizer chambers are filled with a support medium, in this case dry sand. The support medium reduces deformation by a multitude.

Adjustable Mould is another method that uses a flexible formwork skin. It was developed and a first prototype was built under guidance of Dr. Karel Vollers und Ir. Daan Rietbergen at TU Delft (NL), Department Architectural Engineering and Technology led by Prof. Dr.-Ing U. Knaack. And another prototype, that of a horizontal formwork table, was realized by the Danish company "adapa" in Aarlborg, Denmark. The development is intended for thin-walled fiber concrete panels for the façade industry. This system is one of the newer methods to produce double-curved elements. At first, extractable actuators are computer controlled individually to create the shape of the elastic, deformable formwork skin. Then, fiber-reinforced concrete is poured into the formwork surface. After hardening, the mould can be easily peeled of the concrete element and moves into its original shape. Large areas are divided into segments that fit the system size. The data is directly transferred to the mould as is common practice with many other systems. An area with curve radii of up to 40 cm can be produced in less than two minutes. Not yet finalized is the choice of material for the formwork skin. The membrane must be sufficiently flexible to regain its original shape and, at the same time, be of a

1 Sand as a reusable molding material
(Credits: Gramazio & Kohler, Architecture and
Digital Fabrication, ETH Zurich)
2 Vacuum formwork
3 Robot Milling (Credits: Gramazio & Kohler,
Architecture and Digital Fabrication, ETH Zurich)

certain rigidity to resist the pressure load of the material deposited onto and into it. Particularly in the edge sections, strong deformation is to be expected. And, the membrane must retain a smooth surface in spite of any deformation.

Gramazio & Kohler at ETH Zurich, department "Architecture and Digital Fabrication" go one step further. Within the scope of the EU research project "Tailorcrete" they developed a system based on the Adjustable Mould described previously.

Their method involves generating digital wax formwork. A negative form is made from wax, serving as the formwork for in situ concrete. The material choice eliminates limitations in curvature radii, and there is a lot less material and thus energy necessary compared to other, conventional types of formwork (wood). Similar to the Mould, the wax is brought in shape at many individual lifting points with a computer controlled platform. After the wax has cooled and hardened, the form can be taken out. The entire process does not take a lot of time, and the material wax guarantees a smooth surface. After the concreting process is completed the wax can be re-melted and reused.

Within the scope of a current DFG project by the Detmold School for Architecture and Interior Architecture and the TU Darmstadt, vertical form-active wall formwork systems are researched and developed that allow elastic adjustments to the predefined geometric shape for almost vertical formwork surfaces by means of a grid of actuators.

The previously mentioned horizontal formwork tables are simpler than a wall formwork. Only vertical retractable actuators are used to generate the curved surfaces, on which the form-active formwork skin lies loosely. The setup and control of such horizontal form-active formwork tables is fundamentally different from the principle behind angled or vertical wall formwork. Here, the actuators must have a supporting function in addition to imprinting the shape. They must therefore be firmly attached to the formwork skin; an essential factor for setup and control. Currently, a formwork skin is being developed that is expected to exhibit high deformation capability in axial direction combined with good bending resistance against the filling pressure of the concrete. Parameter studies are conducted to examine perforations with regards to their absorption of deformation energy and to determine their mechanical properties.

All of these methods serve the purpose of simplifying the production of a free-form formwork. As Rolf-Dieter Kowalski wrote as early as in 1977:
"...Concrete and formwork are in a perpetual state of mutual dependence. Only the concrete gives the formwork its purpose. But at the same time, only the formwork makes concrete into a usable material..." (Kowalski, 1977)

Based on this quote and the efforts to develop or improve formwork such that free forms are possible, the topic can also be turned on its head. The question then is whether it would not make more sense to create light formwork or even to eliminate its need altogether.

First approaches are to be found around the topic of textile/fabric formwork. This method of constructing is so light that the share of the formwork compared to the used building mass goes to zero. Pioneers in this field are Mark West of CAST at the University of Manitoba and Anne-Mette Manelius at the Danish Technological

Institute of Copenhagen, amongst others. At Detmold School of Architecture and Interior Architecture, the advantages of this formwork method were researched as well. Part of a workshop was the task to fill a "pair of pants" with concrete. Initially ridiculed, this project quickly produced various ideas with potential. There were two basic approaches used; both involved suspending the pants from the waistband and closing off the hem with a wooden board.

For the first approach, the pants were filled while hanging freely. A notable result was that, after being filled with concrete, the fabric of the pants was under so much tensile load that it was formed into the optimum shape of a cylinder. This fact, combined with the still functioning zipper generated the idea of conceptualizing a reusable column formwork. In subsequent steps, various tests were conducted with different skin materials and the application of the zipper.

The second idea arose after a closer examination of the crotch area. After the concrete had hardened, the "pants" were turned upside down. The inseams met in a single point. This realization and the previous idea of a column initiated further development to create a concrete node or a tree-like support. After a suitable cut pattern was created, the concrete formwork was sewn. A framework with tensioning rods was developed to hold the formwork under tension. The tensioning rods were to be used to readjust tension during concreting. The most important finding of this trial was the fundamental importance of the choice of formwork material and seam material. In order to allow the adjustment of tension, the seam and the skin material should exhibit similar deformation behavior.

The second approach also involved concreting the pants in an upright position, however, here, the hems were fixed to the ground. The goal was to stage the pants in a realistic fashion and thus to create a realistic looking form. Due to the pressure load from the poured concrete this proved to be more difficult than anticipated. The initial negative results initiated the challenge to execute the tension-loaded membrane in a way that the final form can be almost controlled. Inspired by undercut anchors available on the market, experimental trials were conducted to study a textile anchoring method for the membrane skin. Different distances between the individual anchor points allowed controlling the characteristic pillow-shaped bulges, and to reduce them to minimum depth. As expected, the shape of the bulge and the arch rise are primarily influenced by the pressure that is generated in between the double membrane but also by the thickness of the final part. For vertical parts, the basic functionality of the hybrid setup, consisting of a support layer and a formwork skin in front of it made of a securely anchored double membrane, could be verified. The experiments included changes in final part thickness, concrete recipe, distance of the counter-anchors and the membrane material.

One significant advantage of a membrane formwork could be the permeability of the membrane material. Excessive mixing water could dissipate through the membrane, which in turn could positively influence the w/z value of the concrete; the expectation being that the concrete becomes more pressure-resistant at the surface. And pore formation could be reduced or even eliminated since air can also dissipate through the membrane.

Another approach deals with eliminating the need for formwork entirely. Counter Crafting is the variant of 3D printing with concrete. The process is still under development but the goal is to design a building on the computer, to send the generated CAD data to the printer, and thus realize the building without any other intermediary steps. In the mind of Prof. Dr. Khoshnevis at USC Viterbi – University of Southern California, in just a few years' time entire buildings could be produced in this fashion within only 24 hours. In addition to time the process would also save material (formwork material) and money, leading to potential reductions in carbon dioxide emissions by approximately 75 % and embedded energy by around 50 %. This fast method would also have the advantage of allowing rapid reconstruction of entire cities after destruction due to natural disasters etc..

It requires a gantry robot larger than the building to produce. The first step is to create a frame (formwork) of up to 10 cm thick from special quick hardening concrete by dispensing successive layers through a suitable nozzle. In order to achieve a smooth surface, trowels are mounted at either side of the nozzle to smoothen the concrete. After the frame is ready, steel-reinforcement can be installed. Now, the frame is filled with regular concrete, and the building stands. Since the nozzle can move freely, not only straight walls but also cupolas or vaults can be made. With this method, apertures must be cut out later.

Gramazio & Kohler from the department "Architecture and Digital Fabrication" and the Institute for Materials (IfB Institut für Baustoffe), both ETH Zurich, examine a digital sliding building method for non-standard concrete elements as part of the running research project "Smart Dynamic Casting".

They examine how geometrically different concrete elements can be digitally produced from plastically shapeable concrete. The focus lies on linking digital fabrication processes with the newest finding from material science. This would allow the development of a robot-based sliding building method that does not require additional formwork.

Outlook

Three established formwork systems are available on the market today, one of which is a pneumatic cushion that can only be used to form equally synclastic surfaces. Research shows that the effort to construct formwork for curved building parts is extremely high. The research can be roughly divided into two categories; one focusing on material evaluation, searching for an environmentally friendly, reusable and easy to process material approach, and the other focusing on new automated manufacturing technologies. Concrete experts are relatively quick in creating new types of concrete to react to new requirements from the formwork industry, such as self-compacting flowable concrete, for example.

In principle, no single type of formwork can be generalized for every building task. But methods that reduce the share of formwork in relation to the weight of the final concrete part or that even eliminate the need for formwork altogether are certainly pointing in the right direction. There are initial approaches to print concrete; however, according to definition these methods are no longer dealing with pure concrete. Mineral printing might be the more appropriate term.

4

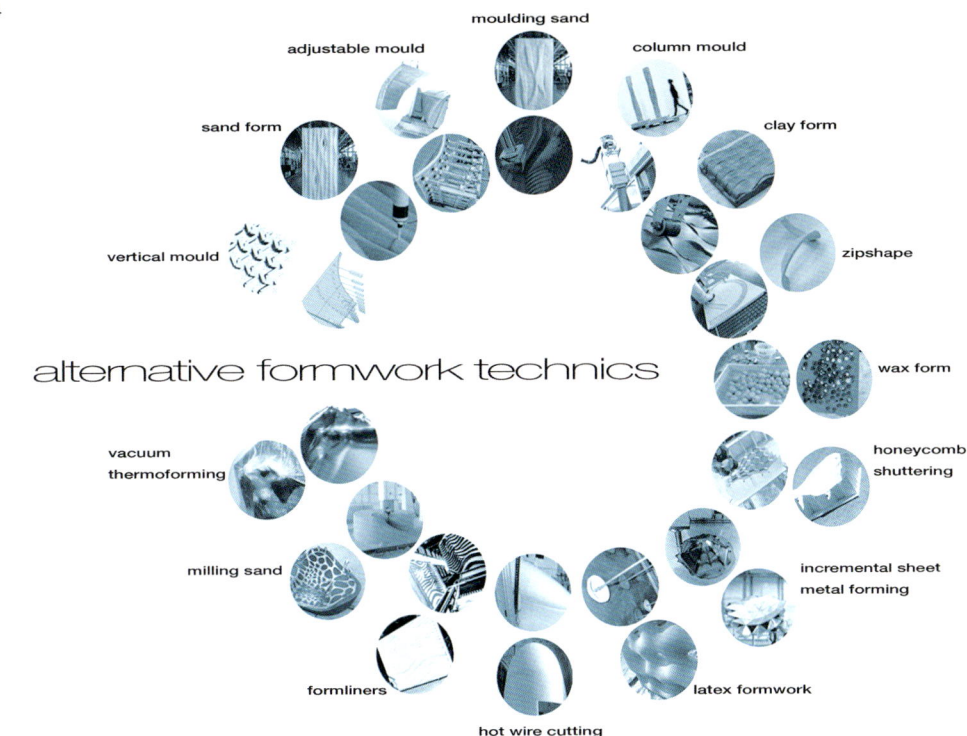

moulding sand

adjustable mould

column mould

sand form

clay form

vertical mould

zipshape

alternative formwork technics

wax form

vacuum
thermoforming

honeycomb
shuttering

milling sand

incremental sheet
metal forming

formliners

latex formwork

hot wire cutting

4 Alternative formwork technics
5 Adjustable mould
6 Column formwork – Zipper
7 Smart dynamic casting (Credits: Gramazio & Kohler,
   Architecture and Digital Fabrication, ETH Zurich)

4. PROJECTS

ARCHITECTURAL CONCRETE FACADE
2012

IMAGINED BY Ducon, RKW architecs
KEYWORDS face concrete facade, cover layer, marble pieces

The entrance area of the "Sevens" on Königsallee in Düsseldorf, Germany, was also developed with the special DUCON Technology, which made it possible to create a concrete roof that cantilevers more than 4.8 m. The panels have dimensions of up to 8 x 3 m with a thickness of 7 cm, which includes a 2 cm thick decorative cover layer with white marble pieces. The panels were sanded and polished to achieve high-quality fair-faced concrete surfaces.

The particular challenge of this project lay in the combination of the fair-faced concrete surface with the fitting accuracy of the large panels with complicated geometries and narrow joints of less than 10 mm.

The construction is carried via a suspension mounting with undercut anchors in the DUCON panel, which meant that the parts had to be very light.

(Credit: Ducon)

COMPOSITE COLUMN
2010

IMAGINED BY Anne-Mette Manelius with assistance from Signe Ulfeldt
KEYWORDS fabric, surface, process, composite, jacket, form, prefabrication, form-tie

Rough wooden formwork boards are still used to create a tactile concrete surface, yet the tectonics of the boards are rarely exploited when they act as the shaping element in rigid formwork systems. Could a prefabricated wood-textile-composite allow the construction of more elaborate concrete geometries and still use very little bracing material as well as saving some concrete?

The Composite Column is a 2 m tall facetted column structure cast in a prefabricated formwork jacket of 'wood plank textile', braced vertically. The approach is based on experience gained from previous experiments: using only material that plays a forming role. Meaning: less bracing, more forming. However, here the intention was to produce a geometrically challenging column with no traces of fabric or bulging. The function of the fabric mould in this case is to act as an embracing pile jacket.

Part of the prototype is the closing mechanism of the jacket, which turns a formwork plank into a form tie and connects it to the vertical braces. Another feature is the 'buttoned waist'; the result of investigating very simple ways to make a different sort of column, an asymmetrical double-column with a specific direction.

COMPRESSION / EXPANSION COLUMN
2013

IMAGINED BY Anne Mette Manelius, Oskar Mannov, Sidsel Petersen, Nora Ødegård, Cuong Tran, and Toke Ridderson.
KEYWORDS fabric, formwork, tectonics,concrete, column, process, form

The use of flexible molds entails a direct formal relationship between the tectonic properties of the formwork structure and the type of concrete used. Thus, there lies potential in developing the details in the formwork construction because they exert technical as well as aesthetic/form-giving influence. On the other hand, this formal consequence means that any mistake becomes visible with equal power; a disadvantage of this building method unless you consider mess-ups as a charming personal note.

Student groups at the 2013 TEK1 workshop at RDAFA came to the same conclusion after preliminary experiments with fabric formed plaster casts. When attempting to literately tailor the fabric formwork, the ruthless character of the poured plaster changed initial intentions to less controlled and little desired folds and bulges. It took a long time to work with intricate principles and it still proved difficult to anticipate and achieve the results they desired. As a result, the students devised a formwork principle that, while simple to construct, still results in sharp, controlled edges as well as soft curving surfaces.

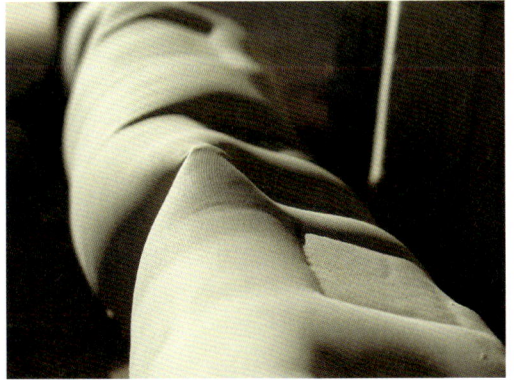

CONCRETE CANOE
2013

IMAGINED BY Ulrich Knaack, Linda Hildebrand, Sascha Hickert and students
KEYWORDS UHPC, folding, glass fiber reinforcement

In June 2013, eight students from the Detmold School for Architecture and Interior Architecture took part in the 14th German Concrete Canoe Regatta in Nuremberg. The students were under guidance of Ulrich Knaack with additional support by Linda Hildebrand and Sascha Hickert.

Even though – or because – none of the participants had much experience, neither in canoeing nor in concrete boats, the team came up with some exciting new ideas. Each student created his own individual design for a concrete canoe. These designs were judged according to specific criteria by an internal jury consisting of Dean Prof. Uta Pottgiesser, Prof. Ulrich Knaack and Prof. Ernst Thevis, and the external member Dr. Marcel Bilow, director of BuckyLab at TU Delft. In a first round of evaluation all designs were presented and examined. Then, the students had the opportunity to present and promote their designs in more detail. The critical aspect was if the canoes would pass the swimming test simulated in 2 m swimming pool.

The jury chose the proposal by Max Ernst and Viktoria Schmunk, who merged their ideas to create one new design. The result was a flat canoe nearly six meters long, of a charcoal color with red seams. The idea was based on a folded canoe, cast entirely flat. Seams between the parts allow folding to generate the shape.
This concept's main challenge was the seams of the fold construction. The best way of construction had to be found by much iteration. The solution was to separate the formwork using an upper and a lower part. The textile fabric was placed in between to act as reinforcement and to connect the pieces. The first problem occurred while pouring the concrete, since the concrete did not flow through the textile fabric. Thus, the layer

Prototype

Reinforcement sew

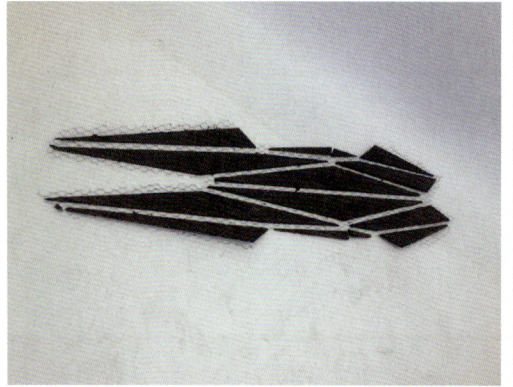
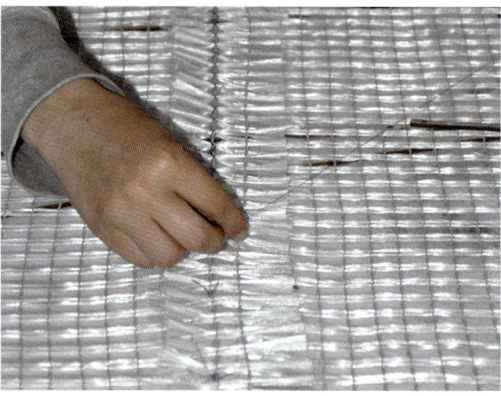

underneath and the layer on top of the fabric had to be cast successively. The different samples did not achieve the desired thickness of 8 mm, causing folding problems. Accordingly, the canoe's different parts had to be poured with the original reinforcement and special concrete mixtures. During this process, it was essential to stretch the textile fabric to prevent it from interfacing with the formwork ground. In some cases, the textile fabric was lifted while concreting. The second method, (paving the concrete over the fabric) was simpler and better, because the textile fabric was exposed on one side. A prototype was built to test the seams and the junctions. The proportions of this model were calculated on a scale of 1:5, but using the original intensity and seam width to be close to the final canoe of 1:1. The formwork was also constructed exactly as it had been planned for the full-size final canoe. That meant that the blueprint was placed on a wooden plate covered with a transparent foil, and plywood elements were fixed with silicone afterwards. After placing the reinforcement, the lower formwork layer was fixed with screws to the upper layer, which consisted of MDF.

All seams for this prototype were 16 mm thick and formed in the flat state. While trying to fold the model, it turned out that the seams were obviously too thick and that the different angles of the elements caused some assembly problems. For visual reasons, in the folded state the seams were to be uniformly 10 mm thick. For all of the above reasons, the formwork for the joints had to be determined again.

Subsequently, experiments were carried out to see how the different compositions would behave, depending on product preferences, namely Nanodur, coarse and finely granulated quartz sand, crushed basalt, black pigments, plasticizer, water proofer, and retarder. When pigments were added, the concrete became very brittle; requiring it to only be used in a lower percentage. Mixing plasticizers with concrete inhibits the bonding properties of the concrete, but it had to flow under the reinforcement, thus high viscosity was essential. The same problem was created by the retarder, even when only a small amount was used. Water proofer had no special effect on the rigidity or setting behavior, but experiments investigating the absorption of the base compound, consisting of cement, sand and crushed stones, demonstrated that water proofer was not really necessary. Finally, to test its critical parts, the top of the canoe was built in the original size.

Concreting

Canoe fold

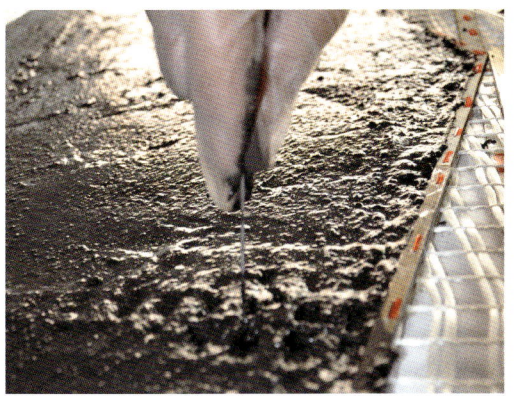

This model and the iterations allowed the team to solve the various problems. The mold of the boat led to different sizes of angles between elements, which meant variable joint spacing. The smaller the angle, the wider the seam distance. The aim was to create a uniform joint, but the new concept did not fit with this aesthetic demand.

On one hand, a gap wider than 1.6 cm did not cause problems for the pointed angles, but on the other hand, the seams in general were not really voluminous and had no grounding. Using an obtuse angle means bridging a high tolerance. Finally, a clearance of 1.0 cm derived from these considerations, although the boards overlapped at the pointed angles and needed to be cut.

For fore and aft, the most pointed areas, we decided to gradate both sides, which meant to gradate the respective sides of the seams and get more space and contact surfaces for the material. Two other seams exhibited the same problems, which we overcame with the same solution; we only had to gradate one of the boards.

The final formwork required several screen printing plates, arranged with a water scale, because any single measuring inaccuracy would resulting in a concrete canoe of a different size. CAD drawings and a laser cutter were used to form the outer boundaries and the joints of the formwork from 3 mm PE foam. In order to allocate the different parts accurately, a plan was plotted and initially mounted on plywood. On top of that, there was a PE film for planning the facing and to provide the opportunity to read the blueprint.

After the first layer of elements was positioned with a fork-lift, the glass fiber reinforcement was stretched out onto it. The next step was to form the second layer, so that the textile fabric could be stretched out and the correct material thickness could be achieved.

For self-compacting, the concrete mix UHPC-Nanodur Dyckerhoff was used. Especially thin constructions can be used to reduce vibration as well the black pigments were perfect match to the canoe's color according to the design. During concreting, it was necessary to lift the fabric in order for the concrete to spread evenly.

On 20 June, the team went to the competition in Nuremberg. Beside the prizes for the sporting competition itself, there were awards for design and construction aspects; as well a T-shirt competition. The jury spent the entire day checking the canoes for accuracy and every single detail mentioned in this report. Then, everyone had to participate in an athletic

Grounting

Finished canoe

competition. For this competition, teams of two had to paddle their canoes over a certain distance and a slalom route.

Our group stood out from the others because it was the only team to consist exclusively of architecture and interior design students, as well as its special presentation and the arrangement of the canoe. The group also drew attention because of our new eponym which attracted a lot of interest all in itself.

Even the most critical person had to admit that our canoe moved very smoothly and safely.

This project has been published several times and was exhibited in Berlin and Düsseldorf at "Architect @ Work" in October. Furthermore, there are many requests for showing the concrete canoe, including several from international sailing and rowing clubs.

Finished canoe

Canoe trip

Maiden voyage

CONCRETE FACADES
2012/2013

IMAGINED BY Sascha Hickert, Linda Hildebrand, Ulrich Knaack and students
KEYWORDS Facades, experimental, face concrete, light

After the success of the student course "Concretable" in the winter semester 2012 / 13, "Concretable 3D – Another brick in the wall" was started under supervision of Sascha Hickert and Linda Hildebrand from the Design and Construction Department of the Detmold School for Architecture and Interior Architecture headed by Prof. Ulrich Knaack. 26 students took part and developed a façade element (120 x 90 cm) with an individual design.

Many ambitious ideas were thought up that led directly to the challenge of the formwork. One group planned to build a facade element that looks like a person trying to free himself from the façade. The first attempt to place a latex layer on a dummy and extract it with vacuum succeeded but did not bring out the desired flowing folds and the textile look. After further attempts, imprints of the arms, head and the upper part of the body were pre-built out of concrete. Then, a vacuum was created under a PVC film and laid over the imprints but crinkles had been created, which led to the use of a gypsum form was taken which served as formwork.

Another team built an element, which was bent on both sides and featured numerous penetrations. Therefore, a two-part formwork box was welded out of steel. Both of the internal sides were covered with EPDM film and the outer sides were closed with an aluminum sandwich panel. Between both layers adjustable spacers for the penetrations were placed opposite each other. During the pouring of the concrete the hollow cells were filled with sand to create counter pressure. The formwork consisted of a timber frame which was spanned with a net of clothesline. Cling film was laid over the frame so that a panel with a little pillow-like case accrued.

Escape attempt

Wax ball element

Another group decided to fill up an air mattress with concrete. In order to avoid the air mattress becoming too heavy a light concrete mix was also used. Egg, shampoo and shaving foam were tried as substitute substances, with shaving foam being used in the end. To absorb the pressure at this point, a gypsum formwork was developed. The air mattress was cut open at the upper end to be filled with concrete.

Another team developed a stick pattern out of individual wooden blocks. The main concern was how to handle the surface of the blocks so they could be easily removed and reused later. The solution was spraying film. Another interesting formwork was built out of silicon to obtain a unique surface. To create the apertures in the silicone, holes were etched into a Styrodur plate with acetone. Others in contrast achieved penetration by using wax balls. Small plastic balls were cut open and filled with wax. The wax semi-spheres were glued onto a screen printing plate. Especially difficult was removing the wax with a hot dryer. The draft for an interior decorative shelf also required complex spatial thinking concerning the Styrodur formwork construction.

The idea of cut open and slightly pressed together paper held the problem of consisting of many component parts, which were supposed to be particularly filigree. Six sample formworks were initially built for the curvatures. After multiple tries of pouring the concrete, the finished curvatures were connected by a closed back board with both endpoints. The group that tried to 'crumple' a steel plate like a piece of paper by trying to jump on the steel and throwing stones at it.

All the final results are exhibited permanently at Detmold School for Architecture and Interior Architecture.

Filigree bows

Concrete air matress

CONCRETE NODE
2014

IMAGINED BY Jens Renneke, Sascha Hickert, Ulrich Knaack
KEYWORDS fabric formwork, freeform, UHPC

Imagine a tree-shaped framework made entirely of concrete.
How can the nodal points of such a structure be made in concrete?
Are textiles a suitable material as formwork material fort his kind of shape?
And how can the joints between the individual posts be design aesthetically?

Concept
The goal of this project was to create a prototype of a nodal point under consideration
of the above mentioned questions.

In order for the prototype to be more than a functionless object its intended use was to
be a frame for a bistro table. Thus, the node was designed with three posts that can carry
a tabletop and three posts functioning as legs. The upper and lower posts were arranged
at an angle of 120° between each other. In addition, the nodal point was not to be merged
from individual segments but rather be generated with a pouring process.

Construction
The formwork consists of a textile made of 100% polyester.
The first step is to determine the cut pattern and to transfer it onto the selected fabric
with the use of a stencil on a 1:1 scale. Then, the individual parts of the formwork are
sewed together.

Since the fabric must be pre-tensioned to balance out the force of the concrete and to maintain of the formwork, a support structure is needed.
The support structure consists of OSB panels and threaded rods, tension belts and wooden slats prevent the formwork of twisting.

The fabric is fixed onto the panels of the support structure with brackets. Hereby, attention must be paid to correctly orientating the textile formwork to prevent contortion within the formwork and resulting creases as these could lead to undercuts and thus to tears or breaks.
The fabric is pre-tensioned by lifting the upper panel with the screw nuts on the threaded rods.

Concrete is poured into the formwork through openings on the upper side. Reinforcement is achieved by adding steel fibres to the concrete.

Conclusion
The textile formwork could be separated from the concrete almost without trace. The fine detailed weave structure of the used fabric as well as the seams of the formwork was modelled onto the surface of the concrete node.

The joining areas of the posts are smooth and do not show any tears or breaks.
Even though the fabric was pre-tensioned it was still too elastic which is why the table legs sagged slightly.

Considering the knowledge gained by creating this prototype it is imaginable that the concrete node could also work on a larger scale.

CONCRETE TABLES
2012

IMAGINED BY Sascha Hickert, Linda Hildebrand, Ulrich Knaack and students
KEYWORDS Bubble desk, surface design, lightweight

In the summer semester 2012, 20 students from Detmold School for Architecture and
Interior Architecture participated in a course offered by the Design and Construction
department.
The task was to answer the following question:
"What can and what cannot be done with concrete? How thin can a concrete panel be
if it has to hold two water tanks placed on top? How can it be built so that two persons
can carry it?"
All participants had to construct a tabletop of approximately 2 x 2 m, which should be
able to be carried by two persons with a container full of water placed on top of it.
This task was carried out with the support and supervision of Linda Hildebrand and
Sascha Hickert, and guided by Prof. i.V. Lutz Artmann. It resulted in eight different tables
with differences in all aspects.

The students generally followed two different approaches to this; one group focused
on the technical characteristics of concrete. The other group focused on the design of
the surface and its outer visual appearance.
The first grouped worked with many different materials and ideas. They examined
Styrodur and expanded clay as well as synthetic balls in terms of their suitability.
One further development was to attach synthetic balls to a basket of screed in a
certain position. In this case the students wanted to create a thick plate which was only
half the weight of a similar plate without these balls. Another approach was to copy the
principle of a coffered ceiling. The construction on the bottom of the desk consisted
of aluminum tubes and small squares of polystyrene which also reduced the weight.
To create a very thin plate, some students came up with the idea to uncover the
construction and achieve the necessary stability by using steel rods as a stress

Technical solutions, uncover construction by using
steel rods

Reducing the weight by creating a coffer cell

component attached to the short sides of the desk. On the construction side there were many different approaches: the panels were, for example, developed with armoring of textile, glass fiber or steel.

The students focusing on the design aspect had a specific idea of the surface appearance. Often they wished for smooth and partially reflecting surface. Those groups also concentrated their experiments on the formwork surface materials. They found out that acrylic is well suited and delivers a smooth and reflecting surface. There were also some ideas to divide the desk into functions, which were marked visually by different surface qualities which could be integrated in some recesses. The students were invited to the concrete factory in Beverungen to cast their tables' tops professionally. The students came up with many interesting approaches for creating a lightweight concrete structure; ideas that can also be used in the building industry. For instance cladding sheets, heat insulating elements or interior constructions are thinkable with such constructions.

It can be expected that some of the solutions will give an impulse for further research. The workshop demonstrated that good technical solutions do not negate good surface design.

Face concrete like a mirror

Exhibition at the Detmolder School for architecture and interior design

Still at work

Group picture

GLIDING WITH WIND AND CONCRETE
2014

IMAGINED BY Manuela Wollmann, Sascha Hickert, Ulrich Knaack
KEYWORDS sailboat, laminated, UHPC, lightweight concrete

The topic of Manuela Wollmann's master thesis under guidance of Sascha Hickert and Prof. Ulrich Knaack was to design, to construct and to build a mock-up of a small sailboat, a sailing dinghy that glides over the water.

But what does gliding in this respect mean? With sailing, gliding is the condition when the hull speed is exceeded by a multitude. The dynamic uplift lifts the bow out of the water, thereby reducing the surface area that is contact with the water and thus the frictional resistance, which in turn makes the boat faster and lets it overcome the bow wave.
To accomplish this, requirements are, on one hand, to sail as upright as possible and to relieve the bow, and on the other hand to carefully consider the weight and the stern end of the hull. It is crucial to keep the weight down, but also to place the centre of gravity as optimally as possible. In the best case this would just a little aft of the centre. The stern should be shaped with a flat, wide floor to achieve optimal conditions.

The design of this project considered the classic lines from the best sailing dinghies of similar size. Another criterion was the transportation medium, a trailer with interior dimensions of 2.95 x 1.20 m, which defined the maximum size for the concrete sailing dinghy.

Working with the material concrete requires formwork. There are two options: either two negative molds, one for the underside of the hull and the other for the deck, or a positive formwork. The success of the first variant would depend on very precise work in putting both parts together, and to be able to connect them in a later process. The second option means that the shuttering would remain inside the boat. Then the idea was developed to use the internal formwork as an additional lifting agent. In addition to air, Styrofoam is an

Concrete lamination process

Ship's mast holder

ideal material for buoyant bodies. The Styrofoam was shaped and then covered with concrete; delivering both, formwork and a lifting body.

After this aspect was resolved the structural integrity had to be taken into consideration. The mast and the centerboard are two points where the hull is exposed to great forces. The mast, a plug mast consisting of a lower and an upper mast, is placed in a mast step and fixed with a closed mast gate. Typically, both brackets are screwed into the hull. However, this is not very trustworthy if you only have a concrete layer of 4 – 5 mm with Styrofoam behind it. Therefore, an additional shaft was constructed. It is made of concrete which can be inserted in the Styrofoam core. Due to its larger surface it offers more primer surface and thus provides a fixed connection between the shaft and the Styrofoam with the help of adhesive power. Strengthening is needed because the ESP can only absorb low pressures up to 100 kPa. The situation of the centerboard is similar. The purpose of the board is not only to avoid leeway but also to re-erect the sailboat if capsized. Herefore, one puts one's entire weight on the end of the centerboard und thus straightens the dinghy up. Accordingly, great forces impact this part of the boat. The centerboard needs a casing with the openings being above the waterline so that water cannot enter the cockpit. With the help of a buffer the centerboard casing guides the centerboard and holds it in position, but the centerboard casing itself must also be supported and fixed in position.

Herefore, vertical frames made of concrete are built, two for the mast, two for the centerboard and one for each point of a horizontal reinforcement. As mast and centerboard are positioned toward the front, the centre of gravity now is in front of the middle. Since it should be further aft, more weight in form of more vertical frames is added to the aft ship. At the same time, these additional vertical frames strengthen the cockpit where the crew usually sits.

The rudder and the mainsheet/the centre padded toe-strap also require brackets. The rudder fits on the transom rudder fitting, which should be immovable because the rudder is subject to great forces. A massive bracket fixed in the last vertical frame is the solution. The brackets for the mainsheet and the centre padded toe-strap are designed according to the same principle.

Exhibition at the Detmolder School for architecture and interior design

The construction process of the Styrofoam core took a lot of time. Styrofoam panels had to be cut into ribs, grinded and glued to the other ribs and vertical frames. At the same time an appropriate concrete composition had to be found. The goal of a concrete sailing dinghy demanded as little weight as possible. This could be achieved with the use of lightweight aggregates and a wall thickness of only a few millimeters, which in turn can be accomplished with the help of textile reinforcement. The concrete should be spreadable, adhere well to the Styrofoam (on vertical and curved surfaces), form a connection with the textile reinforcement and exhibit high pressure and bending tensile strength despite its reduced thickness.

Concreting or laminating was done in two steps, first the bottom side of the hull, then a day later the other side, the deck. We started with a thin layer of concrete, which was deposited onto the Styrofoam by hand, and compacted with gentle tapping. Then an alkali-resistant glass fiber tissue, one more layer of concrete, a layer of glass fiber tissue that was moved to 45 degrees and a final layer of uniform concrete.

Due to the manual work and the concrete composition the finished surface has slight irregularities and a rough surface structure. These are two aspects that might prevent the sailing dinghy from planning, i.e. skimming over the water. Thus, a surface treatment such as grinding is necessary. An additional impregnation is important to prevent the concrete from gaining weight due to absorption when in contact with water.

Finally, when the hull was finished with all mounting arrangements, the brackets for mast, centerboard and rudder could be mounted. For the first time, the concrete sailing dinghy was ready to be rigged.

Styrofoam substructure with UHPC ribs   Section

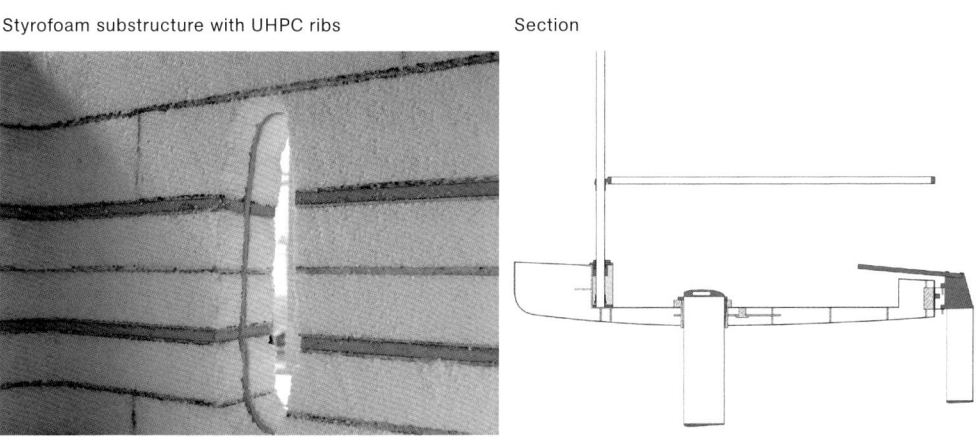

HOUSE IN THE VINEYARD
2011

IMAGINED BY Ducon, TU Kaiserslautern
KEYWORDS self-compacting, micro-reinforces concrete

In 2011 the house in the vineyard was realised as the first of its kind in the vicinity of Mainz, Germany.

The house consists of a total of 7 prefabricated parts, made with the so-called Ducon Technology.

The building parts are only 3 cm thick, even those with a load-bearing function. The individual elements are joined with tongue and groove connections and, at the contact points, glued with DUCON mortar. The roof is a steep roof that transitions seamlessly into the walls.

(Credit: Ducon)

MULTI-FUNCTIONAL CONCRETE BUILDING PARTS

(Thin-walled hypar shell roof structure made of textile-reinforced concrete)
2014

IMAGINED BY Matthias Pahn, TU Kaiserslautern
KEYWORDS fast change materials, integral concrete, thermally activated concrete

Concrete facades and steel-reinforced concrete wall elements open up great potential for energetic use.

By integrating water-bearing pipe registers, entire buildings can be conditioned (heating & cooling) with underfloor, wall and ceiling heating systems. Thermal activation of concrete facades (massive absorber) in combination with efficient heat pumps enables energy generation from renewable energy resources through the facade plane. It is particularly in this area that buffer storage of thermal energy plays an increasingly important role.

Due to the high storage capacity of the material concrete, the load-bearing structure of a building lends itself for heat storage. A targeted integration of phase change materials (PCM) in load-bearing building parts combined with thermal activation can be used to create high performance thermal storage. Compared to regular concrete mass, such PCM concrete parts can offer an increase in storage capacity of 30 to 50 %.

1 Thermally activated concrete sandwich element (M.Pahn, 2012)
2 Phase change material – Paraffin
3 Multi-functional load-bearing concrete building part with integrated PCM

PA COLUMN
2012 – 2013

IMAGINED AND BUILT BY Anne Mette Manelius, Michael Stacey
KEYWORDS fabric, formwork, concrete, for-tie, process, prefabrication

A lot of research is afforded to the form-optimization of beams but columns can be shaped just as well. The formwork jacket is the simplest way of shaping concrete. It can be compared with pouring concrete into one leg of your jeans while holding them upright. The trouser leg will become cylindrical, and the concrete will pack itself upwards and make a vertical, round column. The 'jacket' association derives from the way column formwork is 'dressed' on the reinforcement, and then buttoned or zippered up.
Using the analogy from clothing, could you shape and button up concrete with mechanical means while saving concrete at the same time?

The fabric-formed column designed and constructed for the Prototyping Architecture exhibitions (Nottingham and London, 2013) is an experimental investigation and material prototyping of a lightweight, prefabricated mould, which unfolds for casting onsite. The clustered concrete column is form-optimized for stability and constructed with minimal means.

The three-legged or clustered PA Column is a two-meter high column cast in a prefabricated wood-textile mould in the prototyping hall at the University of Nottingham while the formwork was prepared in Copenhagen.

V-shaped wooden boards are used as forming and bracing elements for the advanced formwork jacket with three 'waists' as opposed to only one for the Composite Column.

PARAPLUIE
2013

IMAGINED BY Ducon
KEYWORDS double-curved, freeform, UHPC

The Parapluie consists of two parts, the roof and the column. The cantilever roof plate with a span of 2 x 3 m and a thickness of only 3 cm is made possible by a combination of the patented DUCON Technology of micro-reinforced high-performance concrete and the membrane load-bearing effect of the construction. The roof plate is mounted onto the slender DUCON column that has a minimum diameter of 12 cm. A steel pipe is positioned inside the concrete column to drain off rain water.

Its shape as well as the slender design of the building parts makes the parapluie appear extremely light and dynamic. The double-curved shape required precise calculations ahead of time, and the formwork was produced with a CNC mill.

(Credit: Ducon)

PRECAST CYLINDRICAL TRC SHELLS
(Lightweight precast cylindrical TRC shell structure)
2013

IMAGINED BY RWTH Aachen, Durapact GmbH
KEYWORDS prefabricated, single-curved, carbon, concrete shells

Due to its non-corrosive reinforcement and high surface quality, textile-reinforced concrete is suitable for thin, high quality prefabricated parts. As part of a current research project at RWTH Aachen, a manufacturing method for single-curved textile-reinforced concrete roof shells was developed and successfully realized in a built project. The chosen reinforcement material was high-strength carbon textile reinforcement, developed at the Institut für Textiltechnik. The realized load-bearing construction consists of a total of 5 concrete shells and serves as roofing for a bicycle parking area.

The shells are only 2 cm thick; the dimensions are 4.40 m long and 2.14 m wide. The arch rise is 50 cm. For architectural reasons the edges of the textile-reinforced concrete shells are chamfered at a 45° angle. The shells rest on steel substructure at four points each. The open span between the support axes is 2.60 m.

Prototypes were used to examine the best manufacturing method and the load-bearing capacity of different variants. Production of the prefabricated textile-reinforced concrete parts was done by the company Durapact GmbH, Düsseldorf-Haan, Germany. Constructing the fine concrete matrix was done in shotcrete by laminating the textile reinforcement layer by layer. Using the formwork repeatedly resulted in economic production of the individual elements in spite of the challenging geometry of the shells.

The realized demonstration structure on one hand allowed an optimisation of the developed manufacturing technique and on the other hand successful testing of building practical processing steps such as safe transportation and simple assembly of the large-scale concrete shells as well decent load bearing behavior verified bt trials at the institute für Massivbau.

1 Assembly of the five shells with the help of a truck-mounted crane and a steel carrying cross
2 Sectional view of the textile-reinforced concrete shell with high-strength reinforcement textile made of carbon fibre
3 Unloading the shells with a forklift
4 Detail of a contact point with holding screws and Elastomer posts
5 Roof structure made of textile-reinforced concrete shells in front of the Institut für Textitechnik of RWTH Aachen (photos: Robert Mehl, Aachen)

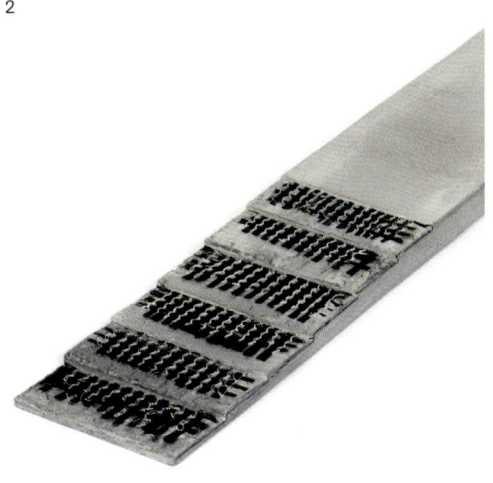

2

3

4

5

ROLLED WALL
2013

IMAGINED BY Carina Kisker, Sascha Hickert, Ulrich Knaack
KEYWORDS core insulation, fabric formwork, density anchoring

The Rolled Wall includes an innovative lightweight formwork system that minimizes the complexity of core insulated double concrete walls formwork and does not require counter formwork panels. For transport the formwork can be rolled up; onsite it is fixed by tension only.

Wooden formwork panels were replaced by textile layers, and the core insulation is done with soft mineral insulation.

The fabric layers are connected to each other to control deformation occurring during the pouring process. Normal steel ties are replaced with nylon strings. The skin, facing the insulation layer, does not let the water from the fresh concrete penetrate.

After curing the outer layer the textile is removed to expose the concrete surface. The textile can also be used as a base layer, if the wall is considered to be plastered. The textile layers can be easily fabricated, requiring only the tensioning frame as an additional part.

The main advantage of this innovative idea is that it provides a lightweight formwork system, which reduces time, manpower and transportation, hence lowers cost. It also provides flexibility in geometric design.

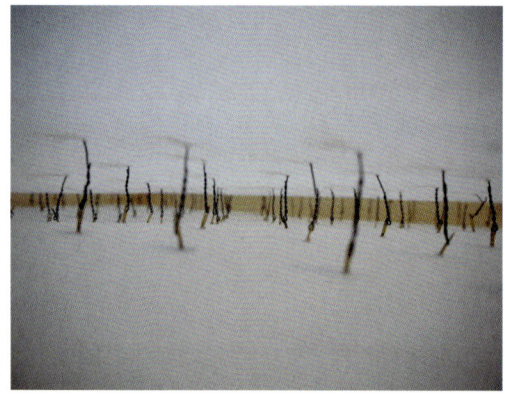

TEXTILE-REINFORCED HYPAR SHELL
(Thin-walled hypar shell roof structure made of textile-reinforced concrete)
2012

IMAGINED BY RWTH Aachen, Durapact GmbH
KEYWORDS shotcrete, carbon, pre-cast shell

A pavilion consisting of four large-sized textile-reinforced concrete shells has recently been realized at RWTH Aachen within the Collaborate Research Centre SFB 532 "Textile Reinforced Concrete" (TRC). A high-strength technical carbon fabric was used as reinforcement material. Due to its high load-bearing capacity, flexibility and non-corroding characteristics the filigree, double-curved TRC shells could be realized with dimensions of 7 x 7 m and a thickness of 6 cm. The pavilion consists of four shells arranged in 2 x 2. At their center the shells are supported by a concrete column. The geometry of the hypar shells results in straight edges yielding a cubic shape for the pavilion with a height of 4 m. The building is encased by a glass facade and will serve as a lecture and exhibition facility[1]. The TRC shells were designed and dimensioned using finite element analysis as well as engineering models for textile-reinforced concrete accounting for material specific effects[2]. The cross sectional strength characteristics of the material have been systematically determined using tensile and bending tests. As a result a total of 12 layers of textile-reinforcement fabrics were chosen, evenly spaced over the cross section of the TRC shell. The developed fabrication concept allowed for an economic realization of the TRC shells due to a repetitious usage of the wooden mold as well as a continuously high fabrication quality. The pre-cast shell elements were mounted on top of the reinforced concrete columns with a truck-mounted crane.

The TRC pavilion demonstrates the strength and application potential of this composite material for shell structures and is intended to inspire architects and engineers for future applications.

1 Schätzke, C.; Joachim, T.; Schneider, H.N.: "Leichte Schalentragwerke aus Textilbeton" in: Beton Bauteile 2012
2 Scholzen, A.; Chudoba, R.; Hegger, J.: "Dünnwandiges Schalentragwerk aus textilbewehrtem Beton: Entwurf, Bemessung und baupraktische Umsetzung" Beton- und Stahlbetonbau, Heft 11, 2012

TRC shells (Credit: Robert Mehl)
Mounting of the TRC shells using a truck-mounted crane (Credit: Peter Winandy, RWTH Aachen)

TILED SLAB STAIRCASE
2008

IMAGINED BY Ducon
KEYWORDS stairs, self-supporting

Typically, a storey-high concrete staircase is realised as a 15 cm thick slab plus steps. In order to minimise the thickness the company DUCON developed a self-supporting staircase in the form of a tiled slab staircase that does not require a separate load-bearing slab and that is only 8 cm thick, reducing the cross section and thereby the weight by up to 60 %.

Until now the staircase has been built in different shades from white to black. Storey-high glass panes are often used for fall protection.

(Credit: Ducon)

TRANSLUCENT CONCRETE
2013

IMAGINED BY AlAmir Mohsen, Ahmed Hafez
KEYWORDS light concrete, integral concrete, glass fibres

Translucent Concrete Panels are a product with which the architect can provide privacy and natural daylight to any architectural space. It could be really useful if used as a material for the main structure as its strength equals that of high-strength concrete. The problem is that to achieve a U value of around 2.15 w/m2k the panel would have to be almost 500 mm thick.

AlAmir Mohsen developed the Translucent Concrete Panel to overcome the drawbacks of the material. Similar to double glazing systems, the Concrete Translucent Panel is made of two panels separated by air or other gas to reduce heat transfer across the building envelope. The first panel is made of translucent concrete where additional fibre reinforcement is added to the concrete mixture in order to maintain small panel thicknesses and overcome structural stress. The second panel is a glass panel to allow the transmittance of light through the concrete panel. The new panel offers a low U value and saves a lot of space compared to thick walls and opaque panels. In addition, different arrangements of the optical fibers offer numerous possibilities for greater freedom of design.

To summarize; concrete has been a trusted material in the construction field for decades, and adding a translucent effect brings it back to life in modern architecture.

WALKING CHROMOSOME
2013

IMAGINED BY Anne-Mette Manelius
KEYWORDS fabric, formwork, concrete, form, process, tailor

What sort of shapes and ornaments can be produced when using a sheet of fabric, hand stitching it closed with metal wire, and shaping and twisting it?

The analogy of casting concrete in a pair of jeans is often used; so what is the translation of producing these jeans in a construction context? There are issues that complicate matters; for example, if reinforcement inside the structure is required you cannot always sew the formwork shut. Therefore, tailoring formwork must involve a different approach to the concept of the seam.

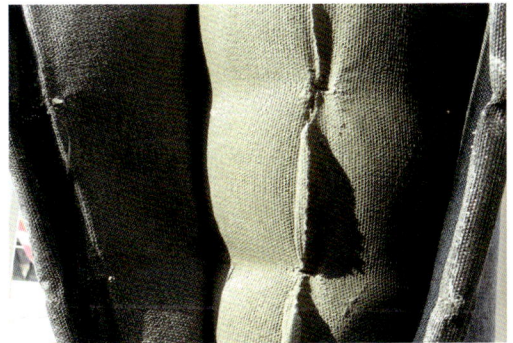
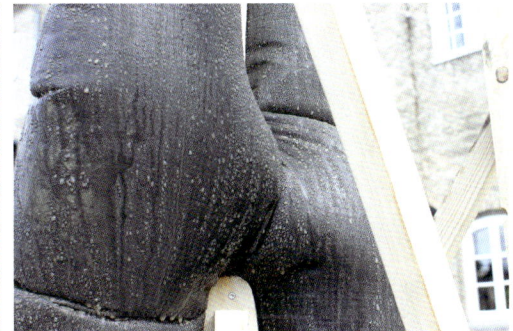

5. PRINCIPLES AND IDEAS

3D CONCRETE PEN
12-12-2013

IMAGINED BY Ulrich Knaack, Linda Hildebrand, Sascha Hickert, Felix Roeder, Rebecca Bach
KEYWORDS 3D concrete, CAM, freeform

The idea behind the 3D concrete pen is that the designer draws the scaled-down structure with a digital pen at home and sends this information to a big pen-like machine outside.

After each completed step the designer clicks 'start' and the big pen 'draws' the 1:1 cubature on site by the use of fast hardening concrete.

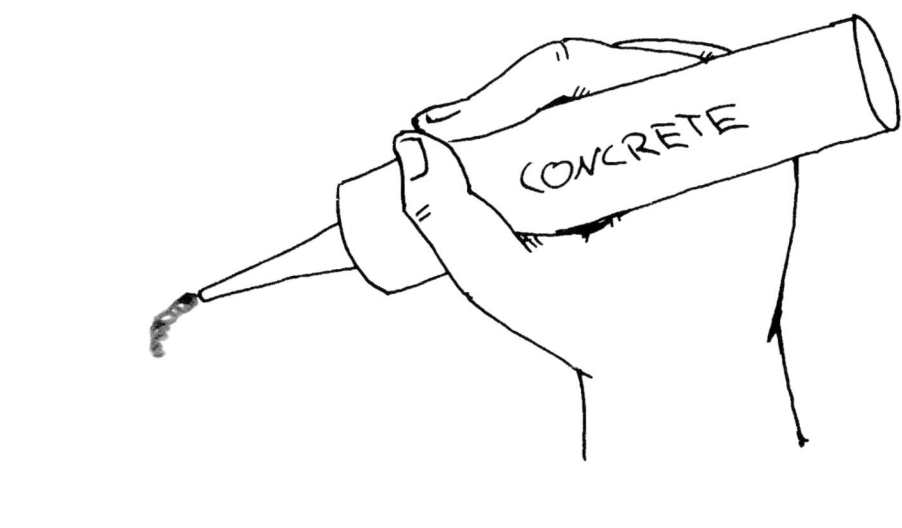

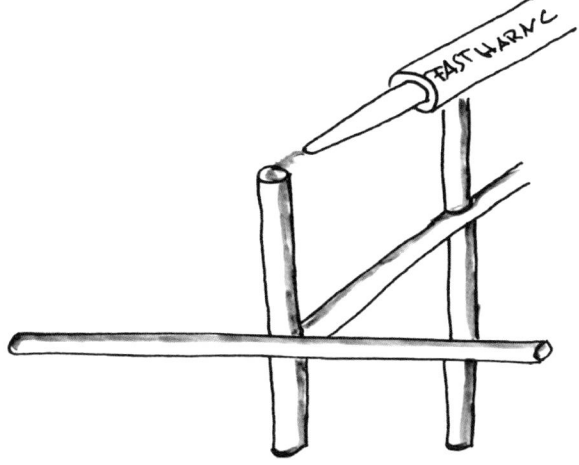

ADAPTIVE CONCRETE WALL
12-12-2013

IMAGINED BY Ulrich Knaack, Linda Hildebrand, Sascha Hickert, Felix Roeder, Rebecca Bach
KEYWORDS thinking skin, technical building equipment

The adaptive concrete wall is a multifunctional system, which includes a huge variety of techniques to influence the room's climate, atmosphere and comfort. Free spaces in the wall can be filled with a variety of additional technical elements.

As there are several spaces for each technical element the assembly process is simplified. The elements are also easy to exchange or disassemble. Technical elements could, for example, be heating and cooling systems, humidification techniques, ventilation systems and glass panes for viewing.

The multi-functionality of this system makes it possible to adapt the technical facilities to the individual needs of the residents.

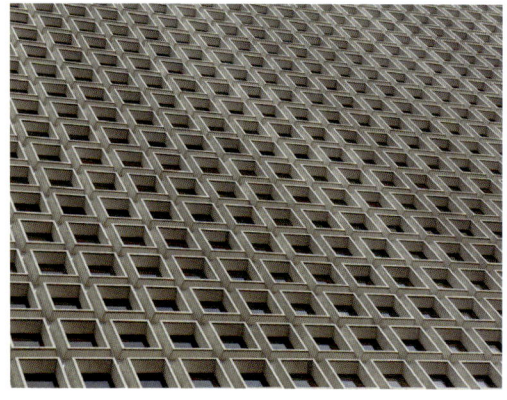

AROMA CONCRETE
12-12-2013

IMAGINED BY Jens Renneke, Reinhard Winzer
KEYWORDS diffusion, air-entrained concrete, fragrance

The idea of aroma concrete is based on diffusing scented air from air-entrained concrete. Plastic containers are inserted into the void in the concrete; these containers can be filled with any type of fragrance. The fragrance evaporates and diffuses through the capillaries and the air voids of the concrete. The larger the amount of fragrance in the voids; the more intense the scent. The reservoirs can be refilled at any time by pouring fragrance into the hollow reinforcement which functions as ducting for the fragrance.

The system could be applied to bathrooms, living rooms etc. instead of using current scent dispensers. The source of the scent is invisible, creating a particular but unconscious sense of the space.

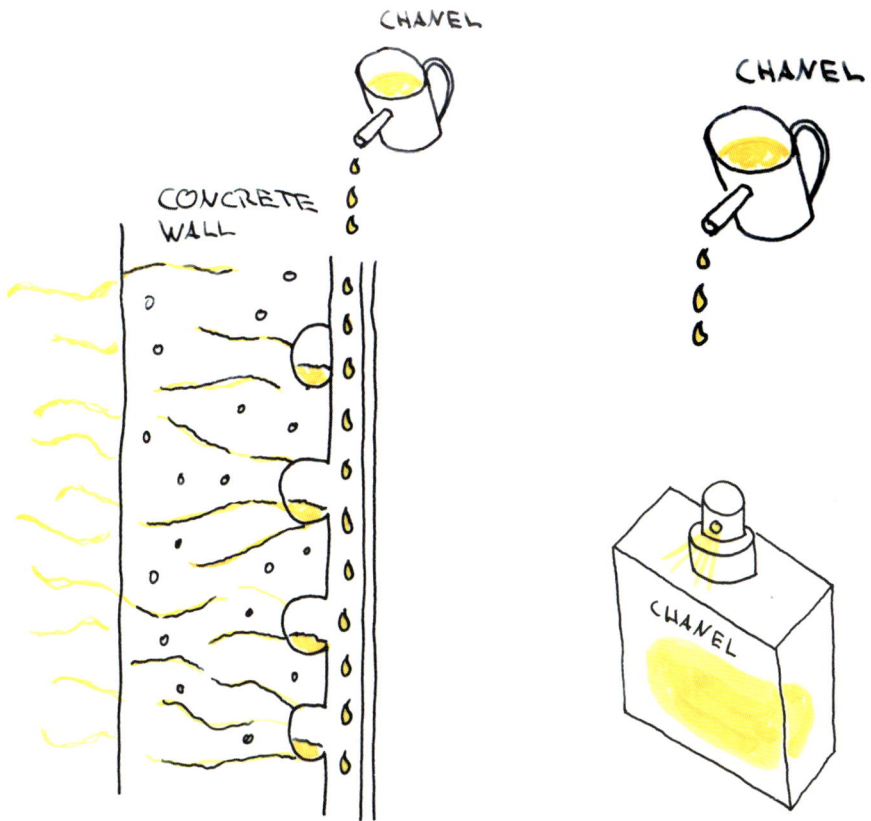

BEYOND MOULD
12-12-2013

IMAGINED BY Ulrich Knaack, Linda Hildebrand, Sascha Hickert, Felix Roeder, Rebecca Bach
KEYWORDS shell formwork, turn around mould, flexible

The main idea behind 'beyond mould' is to use a framed elastomer membrane as the formwork for a structure.

When the concrete is filled in the elastomer deforms in relation to the compressive forces. This leads to a bulge, which forms the moment curve of the structure. After the hardening process this structure can be turned upside down, and the part is perfectly shaped to bear compressive forces on the structure.

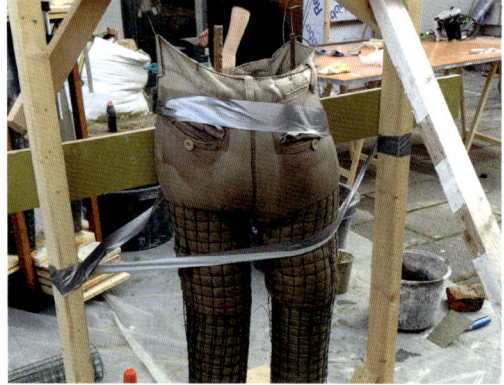
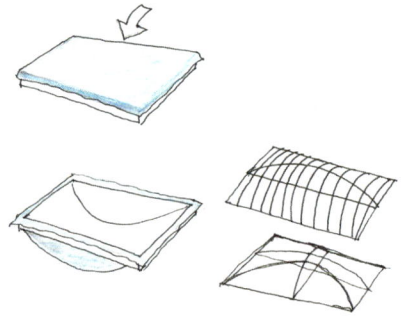

CFE CABLE-FREE ELECTRICITY
12-12-2013

IMAGINED BY Oliver Bach, Nicolas Neidhart
KEYWORDS electricity everywhere, conduction, induction

Today, we all need electricity – ideally everywhere. However, cable ducting constrains the possibilities. We imagined two ways to solve this problem:

A) Electricity through the reinforcement in concrete:
   Electricity is conducted through the reinforcement in concrete walls or, if necessary even concrete ceilings. The positive effects of this system are that milling or mortising into walls or ceilings is no longer required, no cables need to be laid and sockets can be positioned wherever they are needed.
   Different voltages can be achieved by using fewer or more layers of the reinforcement, and the risk of short-circuit faults can be avoided by using a stirrup reinforcement made of hard plastic.

B) Conductive mats arranged at an offset:
   This system is based on two mats, which are arranged at an offset in the depth of a wall. Each mat has a different pole. Thus, one can drill a hole into the wall and place a socket that connects the two mats; the positive and the negative pole. The result is a flexible electricity supply across the entire wall.
   To reduce the risk of a short-circuit fault there has to be a layer of non-conductive fibers in front of those mats.
   This system also enables the possibility to position induction sockets by using a magnet to create an electric field.

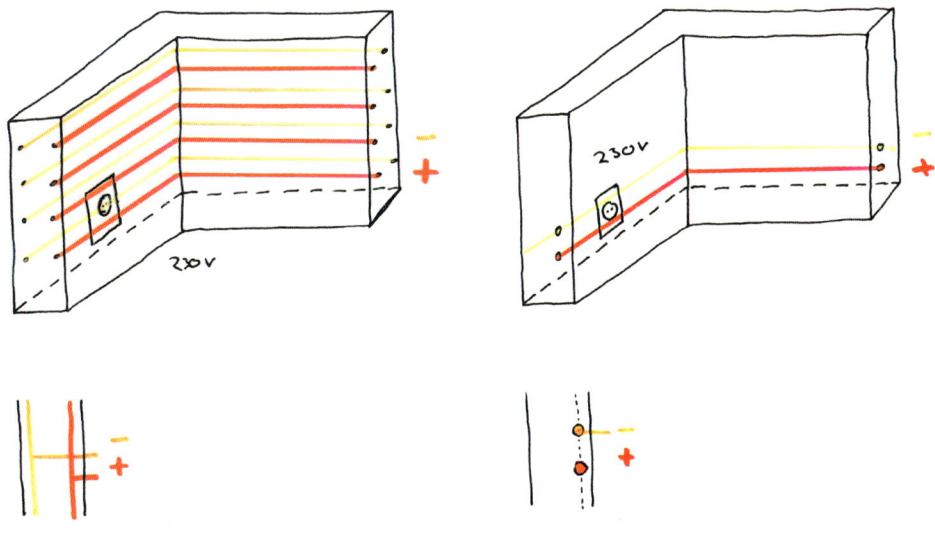

CONCRETE CANNON
12-12-2013

IMAGINED BY Tim Winter, Willy Schwenzitzki
KEYWORDS repair, surface, shooting, far distances

The concrete cannon solves the problem of damaged areas in a concrete structure that are hard to reach.

Fresh concrete to repair those damages is formed into ball shapes, which are covered with an elastomeric coating. The cannon can shoot these concrete balls toward the target, the damaged area of the concrete structure. The coating opens just before the bullet hits the target; so that the concrete is able to stick to the damaged structure. The coating could also be used as additional protection against weather or other damaging influences.

The coating could open up in several ways: either by bursting open the elastomer with a time fuse or by melting it with help of the heat that is caused by the hydration process of the concrete.

The concrete bullets would have to rather small to achieve a smooth surface.

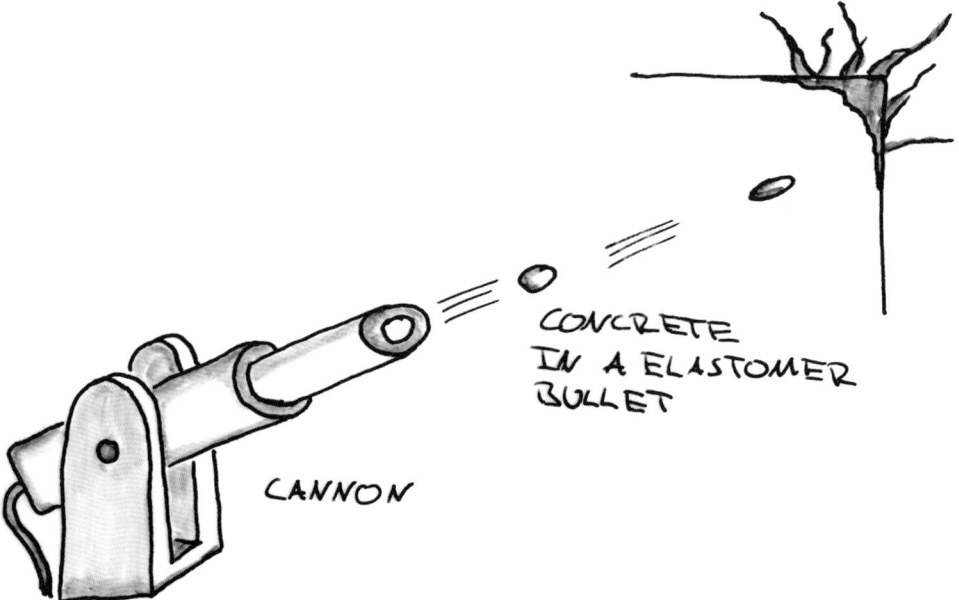

CONCRETE IN A ELASTOMER BULLET

CANNON

CONCRETE DAM
12-12-2013

IMAGINED BY Marius Schmidt, Julian Lianarachchi
KEYWORDS flooding, dam, bridge

The concrete dam is an adaptable safety wall for flood areas that provides openness and penetrability when not in use while safeguarding the living environment during flooding.

It consists of parallel arranged rectangular concrete structures. These rectangle blocks are positioned near the sea or river shores that are known to be endangered to overflow. In case of imminent flooding, the blocks can be turned by 90 degrees to form a continuous dam and hold back the water.

In the flood safety arrangement they serve as a walkable bridge-like structure, potentially used to survey damages.

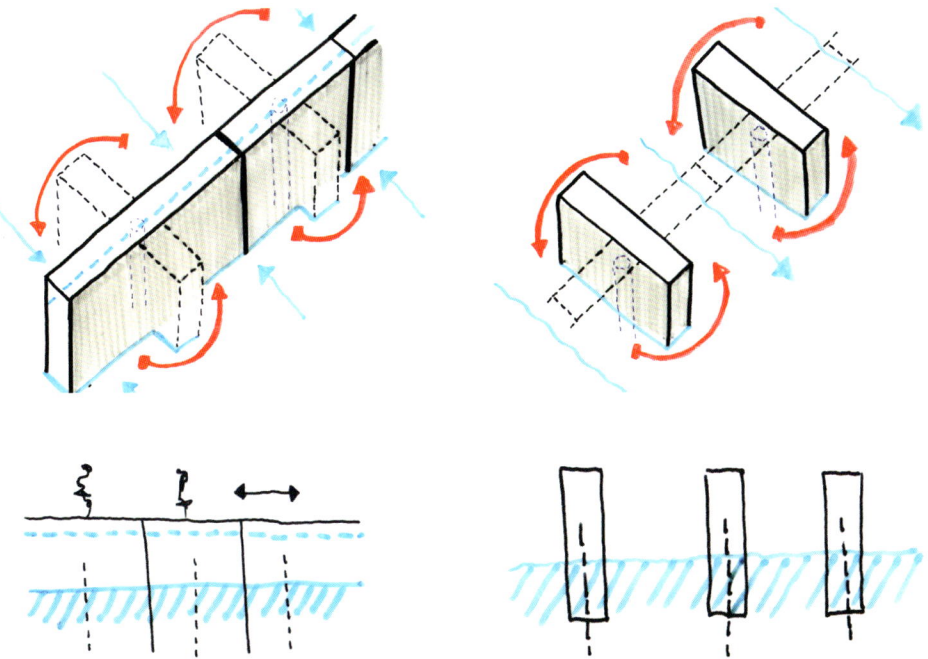

CONCRETE EXTRUSION
12-12-2013

IMAGINED BY Ulrich Knaack, Sascha Hickert, Linda Hildebrand
KEYWORDS injection-mold, extruded concrete, 3D printing, rapid

The production of PVC and other plastic strands with the help of extrusion processes is common practice. Concrete can also be extruded with similar processes.

Fresh concrete is filled into the machine and runs through particular processes, which form and partially harden the concrete in a manner that a concrete strand can be pressed out at the end.

The next step of concrete extrusion (extrusion 2.0) is to think of it in three dimensional terms: the final product does not have to be a straight-lined strand but a free-formed one. The x direction and shape of the strand is still done with extrusion. But the y and z directions have to be done by moving the machine structure.

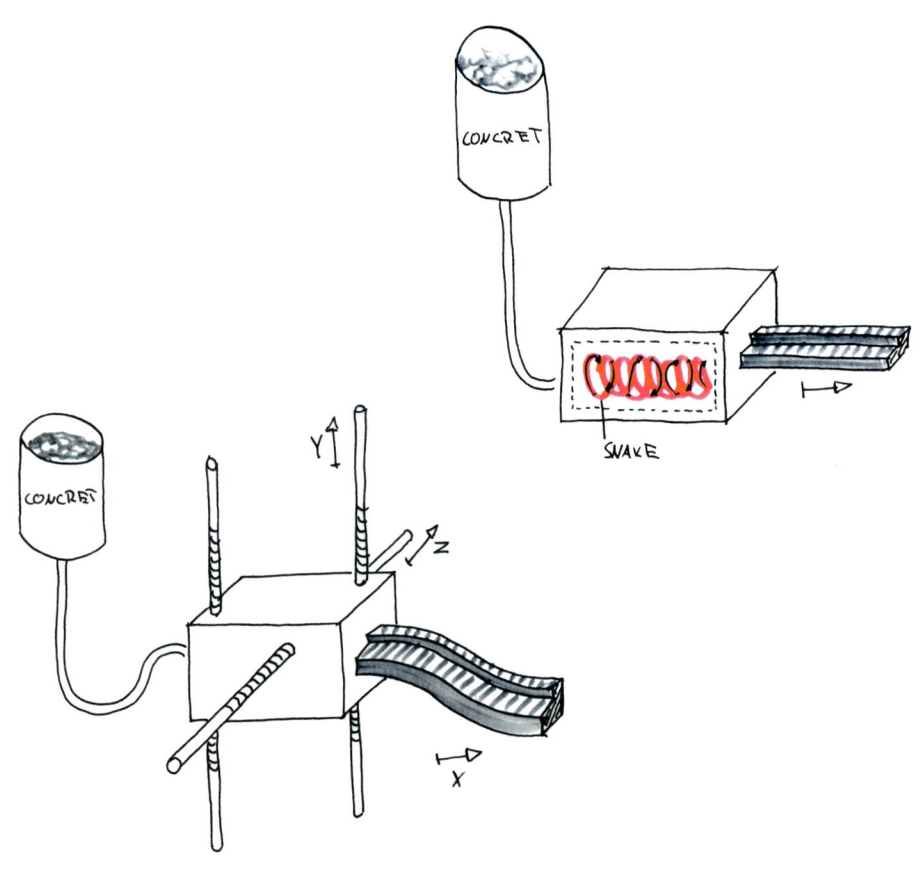

CONCRETE JET PRIMER
12-12-2013

IMAGINED BY Ulrich Knaack, Linda Hildebrand, Sascha Hickert, Felix Roeder, Rebecca Bach
KEYWORDS freeform, CAM, powder mould

There is no need for formwork to mould (free-formed) concrete.

Here, concrete powder is deposited on a plate. Then an automatic squeegee pushes those parts away that are not needed for the final form. After this step water is poured over the powder and the concrete can start the hardening process.

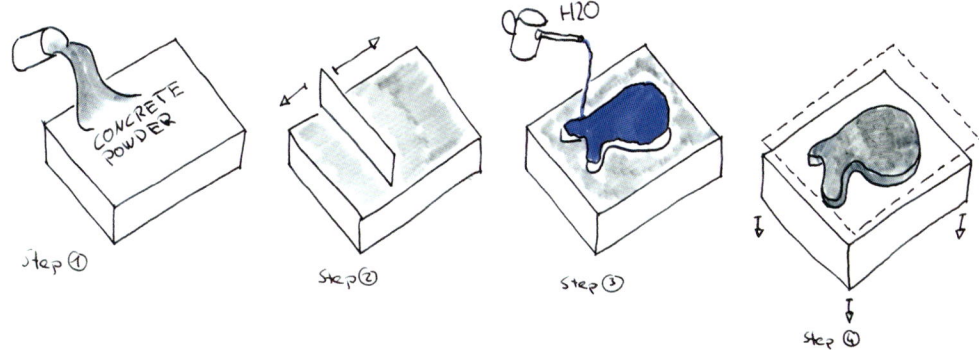

DOUBLE LAYERED ZIPPING COLUMN
12-12-2013

IMAGINED BY Ulrich Knaack, Linda Hildebrand, Sascha Hickert, Felix Roeder, Rebecca Bach
KEYWORDS zipper, stacking, concrete node, in-situ, fabric formwork

The double layered zipping column is a formwork system to form a hollow column.

This system consists of two layers; the outer layer is a textile furnished with a zipper to offer reusability for further columns. The inner layer is made of a textile filled with water, to form an inner water column, which displaces the fresh concrete in order to form the hollow core. With a special connecting dot, several columns could be stacked on top of each other.

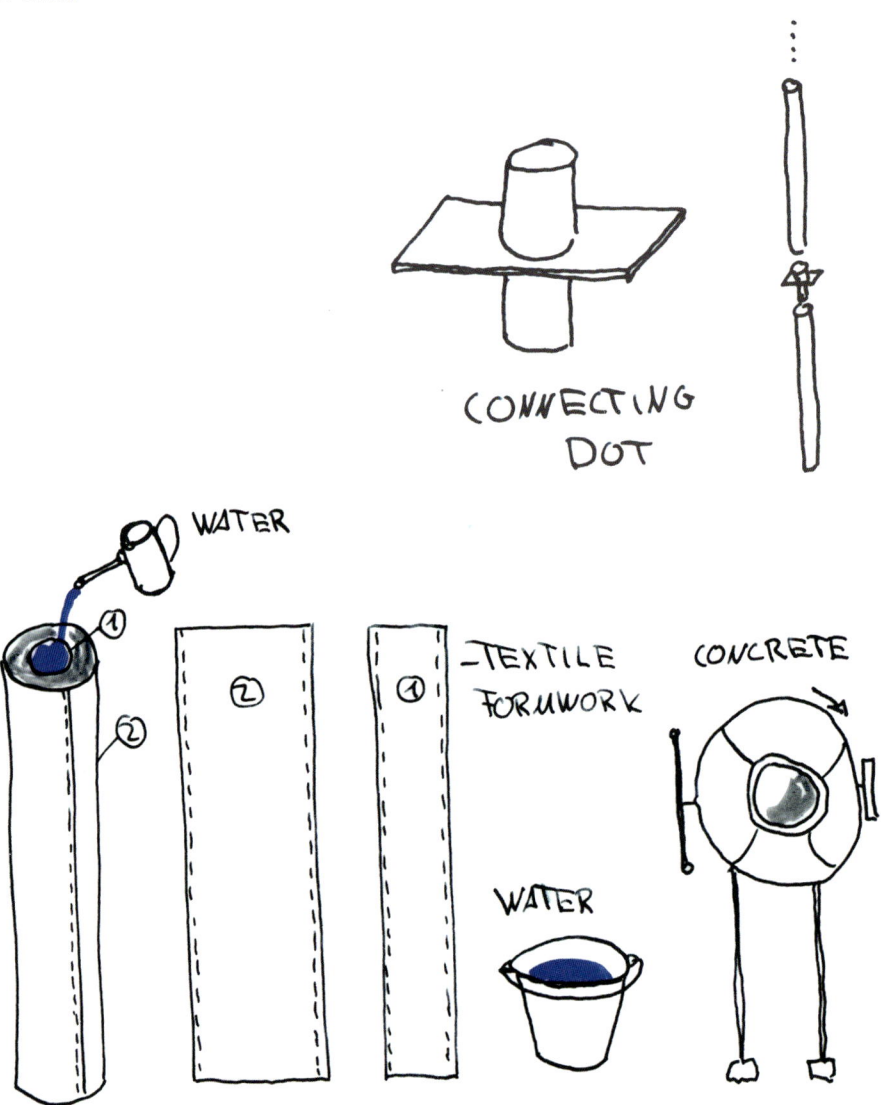

CONNECTING DOT

WATER

① ②

TEXTILE FORMWORK

CONCRETE

WATER

FORCE DRIVEN TEXTILE FREEFORM
12-12-2013

IMAGINED BY Ulrich Knaack, Linda Hildebrand, Sascha Hickert, Felix Roeder, Rebecca Bach
KEYWORDS stilts, form follows force, free-form

This formwork consists of a textile to create an uncomplicated formwork for a free-formed object.

The textile is clamped in a frame, and can be additionally fixed to other parts of the construction to create many different shapes. The textile deforms in relation to the compressive forces of the concrete. Thus, the form follows force.

This system offers new possibilities for free-formed structures and facilitates formwork fabrication processes, because there are no complex negative forms needed anymore.

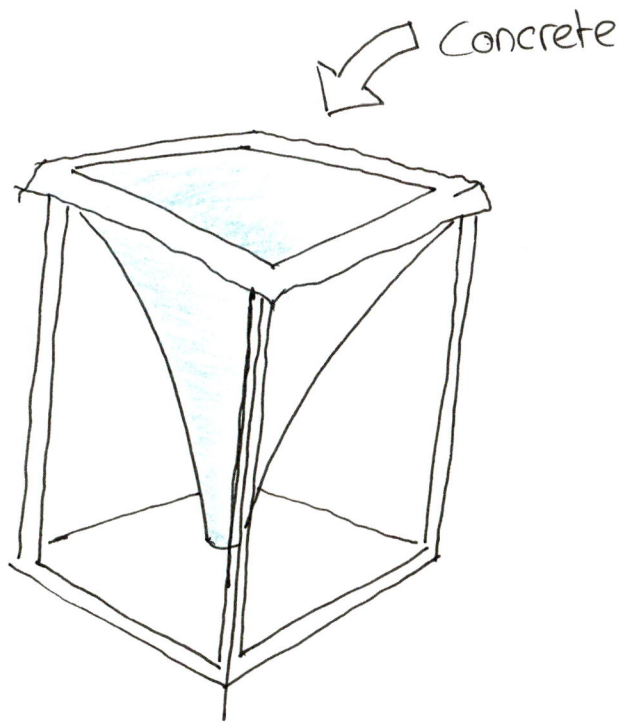

Concrete

GROWING CONCRETE
12-12-2013

IMAGINED BY Ulrich Knaack, Linda Hildebrand, Sascha Hickert, Felix Roeder, Rebecca Bach
KEYWORDS fine dust eliminator, CO_2 concrete, highway, noise protection

Concrete is mostly made of natural materials, so why shouldn't it grow?

Using fine dust, for example near motorways, and water from rain, the only ingredient missing to produce concrete is cement. Hence, while it's raining a small vehicle could drive along a concrete wall that could do with an additional layer, whirl up the fine dust from the ground and spray cement on it – the wall grows. In addition to the growing process this principle would reduce the fine dust content near motorways.

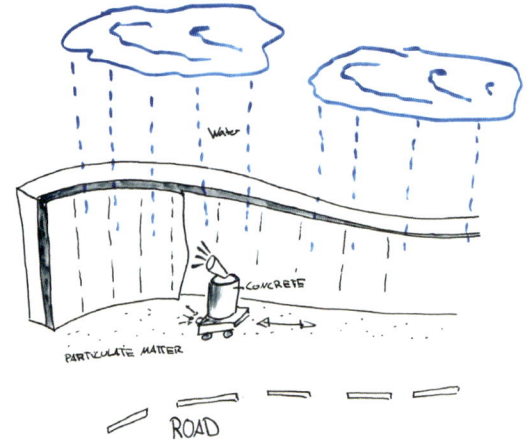

HOME BAKED CONCRETE
15-02-2014

IMAGINED BY Thomas Henriksen
KEYWORDS concrete elements, houses, flexibility, user defined shapes

Imagine it was possible to build your dream house by creating the shape with an app. The app automatically orders the elements necessary to create the shape and the concrete structure. The elements are light-weight and can be easily handled, and no steel reinforcement is necessary because the material properties of the concrete differ depending on the structural and thermal application. The elements are sandwich elements with vacuum insulation and offer the flexibility to add windows and openings where necessary.
The elements arrive as a flat-pack; they come with instructions and can be connected like puzzle pieces. All installations are integrated in the elements and are easily connected. The interfaces between the elements are connected with a premixed concrete material just like a cake mix where the only thing you need to add is water.

The light-weight concrete elements are insulated with vacuum insulation

Openings can be integrated in the elements where necessary

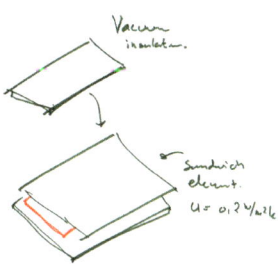

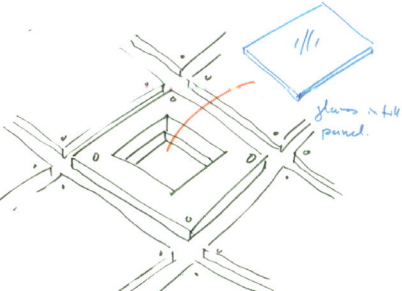

The elements are easily setup and connected onsite, similar to flat-pack furniture

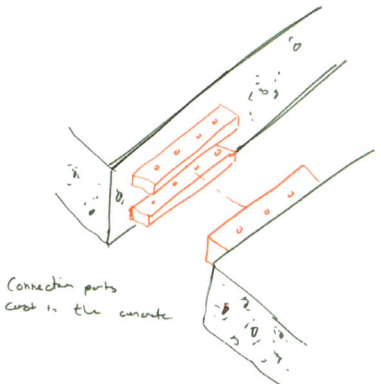

HUMIDITY DRIVEN GROUND WATER COOLING
12-12-2013

IMAGINED BY Ulrich Knaack, Linda Hildebrand, Sascha Hickert, Felix Roeder, Rebecca Bach
KEYWORDS ground water, humid walls, cooling

This system exploits the advantage of the cooling performance of a wet wall for the cooling process in summer.

The building's basement is built in an excavation with circulating ground water, without any waterproofing foils. Thus, the walls soak up the water, which causes a cooling effect inside the building. To control the humidity in the wall and prevent molding, there are bulkheads that can be closed in winter to dry out the basement.

WINTER — Dry Basement — no water

SUMMER — Basement in Ground Water — cooling with wet wall

JUMP AWAY – LIGHT AND HEAVY DUTY
12-12-2013

IMAGINED BY Rütt Schulz-Matthiesen, Manuela Wollmann
KEYWORDS crash barriers, self-evasive, energy absorbent, bending

Jump Away is a safety system of self-evasive crash barriers. They absorb the energy of a vehicle crashing into them by bending backwards; ideally stopping the vehicle completely. These barriers are fixed to the ground with springs or elastic metal strips, which allow the system to bend repeatedly.

In addition, these barriers have a special convex shape in the lower area. Thus, the initial deceleration of the crashing vehicle comes from the convexity of the barrier, and the second one results from the vehicle itself bouncing backward.

Ideally, the Jump Away should be available in two variants. A lighter, less strong model for parking garages, and a heavy-duty one for streets and highways that can take a lot more load due to higher driving speeds.

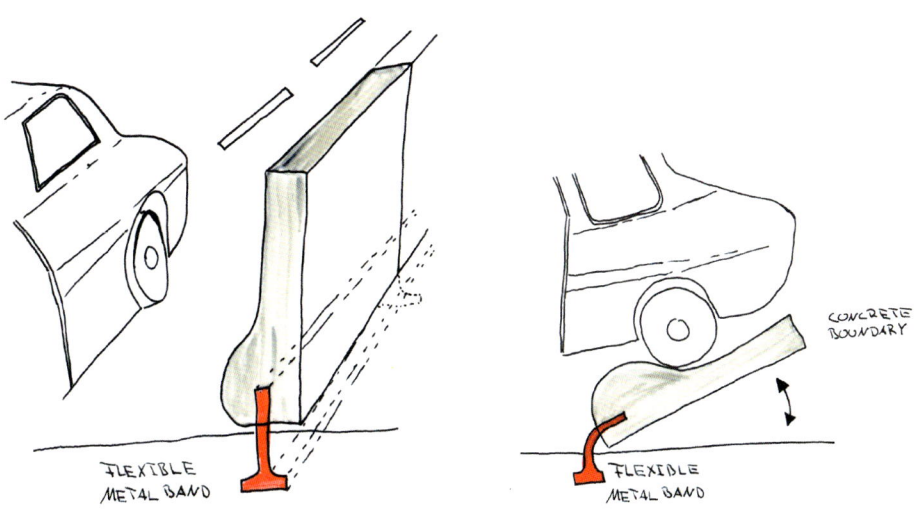

FLEXIBLE METAL BAND

FLEXIBLE METAL BAND

CONCRETE BOUNDARY

MULTI CONTOUR CRAFTER
12-12-2013

IMAGINED BY Ulrich Knaack, Linda Hildebrand, Sascha Hickert, Felix Roeder, Rebecca Bach
KEYWORDS 3D printing, CAM, free-form, contour crafting, rapid

A three dimensional printing system is used to create a free-formed concrete structure. The concrete is applied through nozzles that dispens it on a horizontally orientated plate, layer by layer. Each layer can be shaped in any free form due to the different directions of motion of this system. The nozzles are moving in Y direction, while the whole plate is moving in X direction. So each layer can be formed individually and build up a unique shape very efficiently.

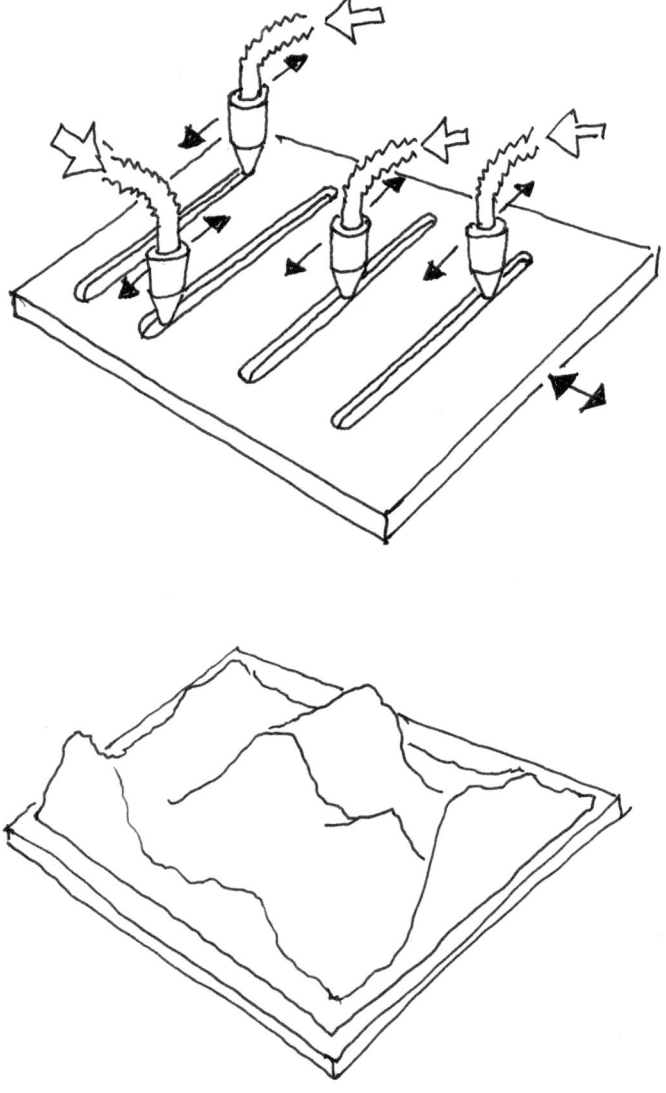

NON-EULER CONCRETE COLUMN NECC
12-12-2013

IMAGINED BY Ulrich Knaack, Sascha Hickert, Linda Hildebrand
KEYWORDS fabric formwork, buckling, form follows function

The Euler cases demonstrate the weak points of a column, those areas where the column is prone to bend and/or break when forces are applied. So the Non-Euler Concrete Column gets a different shape to counteract this problem.

With the help of a particular cut and sewing pattern to create a textile formwork, the column is formed with a bellied shape in the center section, which prevents it from bending and breaking.

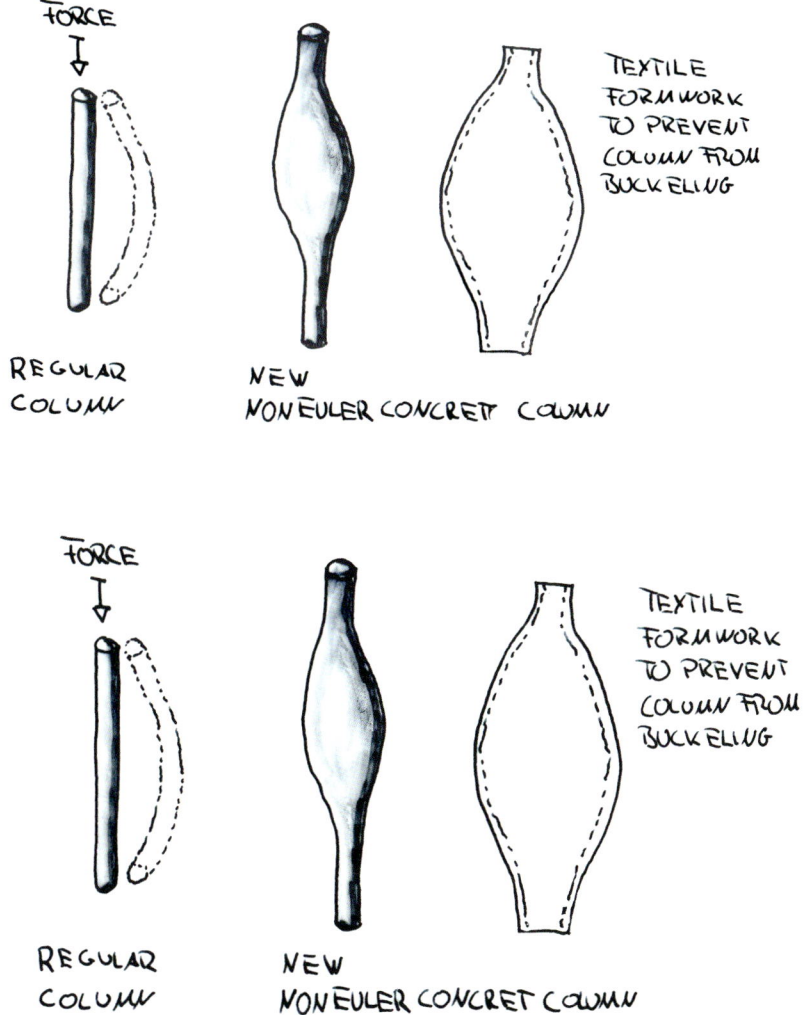

PNEUMATICALLY GENERATED RIBBED SLAB
12-12-2013

IMAGINED BY Ulrich Knaack, Linda Hildebrand, Sascha Hickert, Felix Roeder, Rebecca Bach
KEYWORDS bubble desk, lighter slab, pneumatic, reusable

The principle of the pneumatically generated ribbed slab is to reduce the weight of this structural part by creating air voids within the slab.

In contrast to the usual ways of creating a ribbed slab structure, this system uses different formwork parts. The sides and the top parts are made of a usual formwork to create smooth surfaces. The part on the bottom side is perforated and foreseen with an EPDM foil attached to the underside.

After fresh concrete is filled into the mold, the foil is inflated. It expands within the concrete, hence displacing the concrete so that air voids are formed.

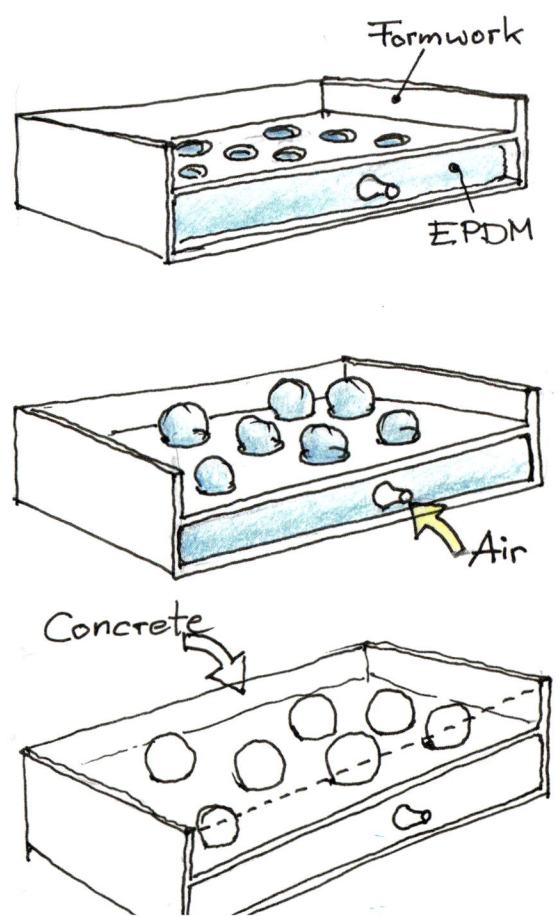

SHINING CONCRETE
12-12-2013

IMAGINED BY Linda Hildebrand, Jens Renneke, Rebecca Bach
KEYWORDS light by loads

Shining Concrete is a system for cantilever structures. It makes the surface light up with integrated LEDs when the structure is subjected to overload. If the entire platform shines, this indicates an overload, which should be removed to prevent the cantilever from being damaged or breaking.

This system works by using bimetal as reinforcement material. Bimetals are composed of two metal plates that bend at different rates when they are loaded. Hence friction is created which produces electricity. This electricity can be used for the LEDs to shine. Each LED is connected to a certain region within the cantilever. Therefore it is possible to see, where exactly the concrete has to bear the loads at any given moment.
The system can be used for viewing platforms or other kinds of cantilevers. It could be used as a fun feature for people to walk on and create different patterns but it also helps to ensure their safety by giving immediate indication of a potentially dangerous situation due to overload.

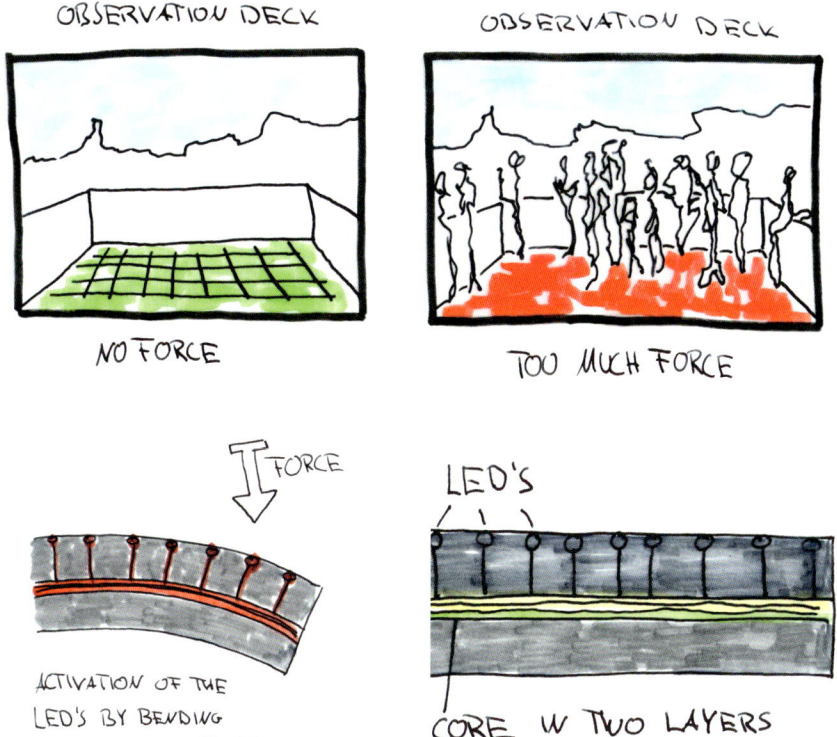

OBSERVATION DECK

NO FORCE

OBSERVATION DECK

TOO MUCH FORCE

FORCE

ACTIVATION OF THE LED'S BY BENDING OF THE CORE LAYERS

LED'S

CORE W TWO LAYERS

SHRED YOUR HOUSE AND STORE ENERGY
12-12-2013

IMAGINED BY Ulrich Knaack, Linda Hildebrand, Sascha Hickert, Felix Roeder, Rebecca Bach
KEYWORDS demolition, sun collector, storage, decentralized landfill

This system works by recycling concrete – waste utilized for energy storage.

After the demolition process the concrete parts of the building are collected and embedded in the ground, where a new building is to be built. This embedded concrete waste works similar to a long-term storage for energy in the form of heat. The thermal mass of the concrete is used as the storage medium. The concrete waste can be warmed up in summer by a special heating panel system, for example on the roof, and gives off the heat in winter, when it is needed. The basement of the house serves as a buffer zone to control the thermal impact on the rest of the building.

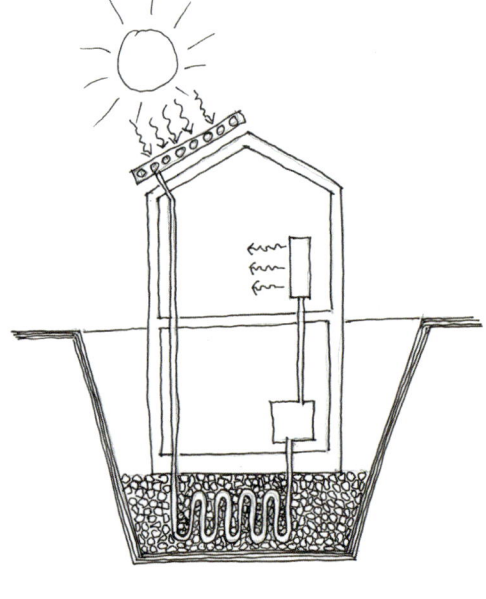

TALKING DOTS
12-12-2013

IMAGINED BY Jan Baumgartner, Achim Hoffmann-Brüning
KEYWORDS transmission, conduction, information

Talking Dots transmit a signal, which is transmitted inside the concrete toward a receiver; for example a mobile phone or similar device to report a special situation, for example, a damaged bridge due to overload.

The talking dots are positioned between every reinforcing steel, at the position of a stirrup reinforcement. Electricity is conducted through the reinforcement. Thus, a different solution for the stirrup reinforcement is needed than the usual welding seams. Plastic clamps, for example, could be used at this point.

If the bridge or any other concrete structure equipped with this system is subjected to an overload, those plastic connections break. With the result that the reinforcement steels get in contact. This contact in turn causes a signal to be transmitted, which informs someone about possible damage to the structure. The system could also be combined with some kind of camera, which takes a picture of the vehicle that ignored the bridge restrictions by crossing it with too much weight.

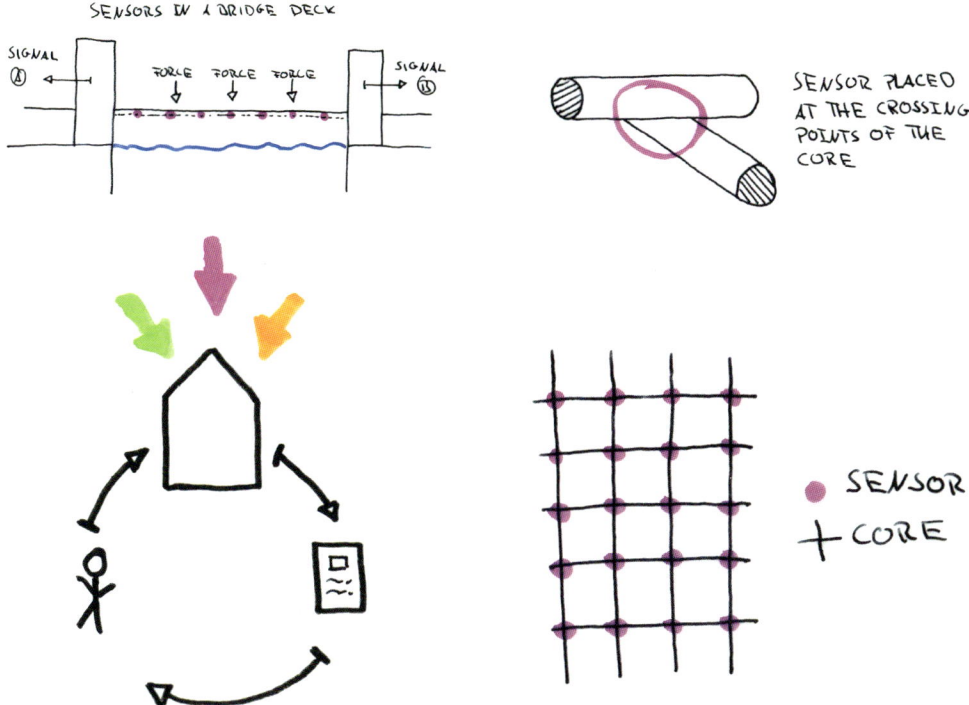

TUBE FORMWORK
12-12-2013

IMAGINED BY Ulrich Knaack, Linda Hildebrand, Sascha Hickert, Felix Roeder, Rebecca Bach
KEYWORDS formwork from the roll, zipper, reusable, endless

A flexible tube can be used as a formwork system. The tube is furled on a long role, comparable to a fire hose. Such tubes can be manufactured with different diameters.

They just need to be unrolled and cut to the desired length. A zipper along the entire length of the tube ensures the reusability of this system.

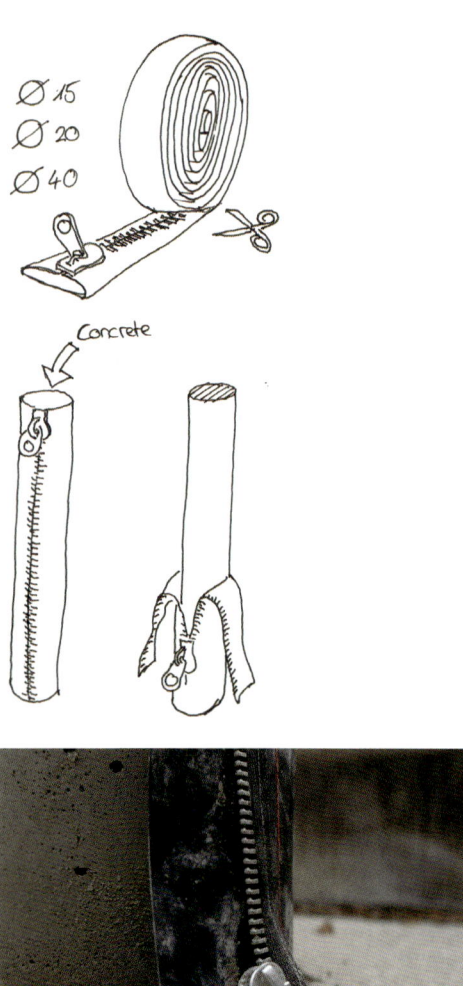

WATER MOULD
12-12-2013

IMAGINED BY Ulrich Knaack, Linda Hildebrand, Sascha Hickert, Felix Roeder, Rebecca Bach
KEYWORDS counter pressure, semi-double shell mould

For this principle, a double layered system is used to realize a slimmer profile than the ones which are created in 'beyond mould'.

The bottom layer is still made of an elastomer, which is clamped on a frame and deforms due to the compressing forces of the fresh concrete. The upper layer is also made of a clamped elastomer. After the concrete is filled in the bottom layer, the upper layer gets filled with water to displace the concrete and get a slimmer profile, which still has the perfect shape to bear compressive forces after turning it upside down.

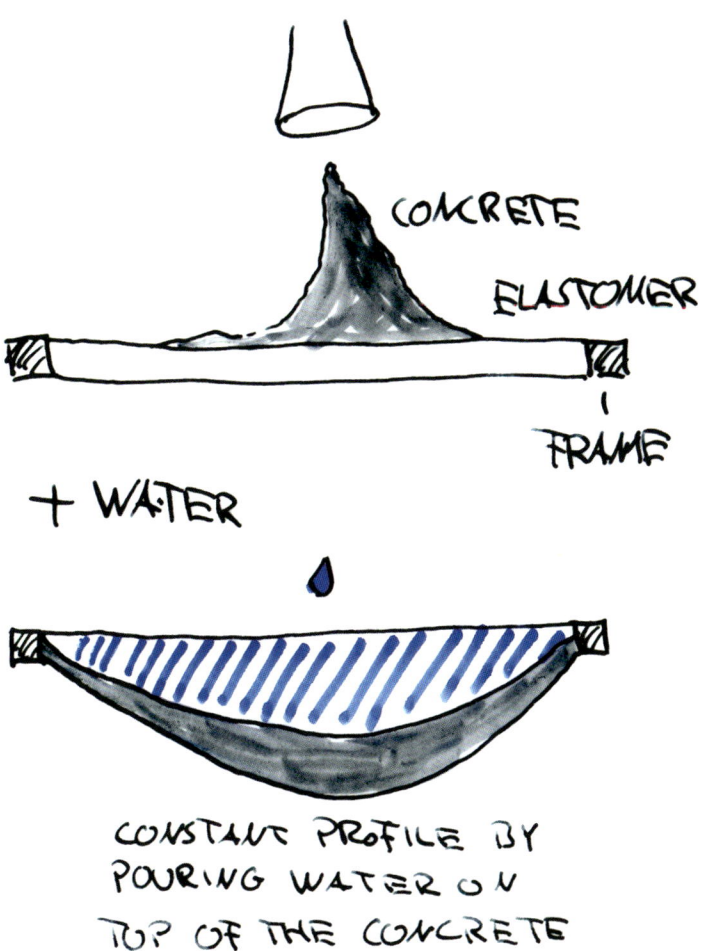

6. PERSPECTIVE

THE FUTURE WILL BE LIGHT AND COMPLEX OR HEAVY AND SIMPLE

When examining the global trends in the building industry in general and those of concrete constructions in specific, five topics can be isolated that will have a significant impact on future developments for the material concrete.

ECOLOGY

There is no need for lengthy discussions in order to justify the issues of ecology and sustainability of building materials. However, concrete does play a special role since 3 – 5 % of the global CO_2 emissions are caused by producing cement for concrete. And considering the performance capability of the material, in a facade for example, and comparing it to that of other facade materials such as timber, steel, aluminum and glass, the fact is that the performance of concrete in terms of the energy needed to produce it is lower than that of other materials. Thus, concrete rates worse in according evaluations.

The goal must be to improve the performance capability of concrete to enable more cost-efficient constructions – but more about that later.

One approach is to improve the possibilities for the use of the material after its current use: today, concrete is fed into a down-cycling process once the end of life of the original product is reached. It is interesting to see how much improved and more efficient the technology of deconstructing the composite material concrete has already become, and how well we are already able to machine process concrete building parts back into metal and mineral materials – even if the mineral share cannot be reused at the same functional level as steel, for example, which can be melted and thus reused. If we were to develop a material such as concrete in current days, we would probably not conceive it as an inseparable composite but rather concentrate on solutions that are easier to disassemble.

Thus, the objective for an alternative concrete material would be a repeatable solution with a separable connection of the mineral components – a nice goal.

EQUAL TENSILE / PRESSURE STRENGTH AND CONTROLLED STRUCTURAL STRENGTH

The interesting aspect of the material steel from a constructive viewpoint – and certainly the reason for the material's success – is its potential to absorb both tensile and pressure loads. It allows us to develop constructions with undirected force distribution and maximum load concentration in a very small area.

Concrete can deflect pressure loads well, but it does require additional reinforcement to bear tensile loads. Therefore the high number of approaches to reinforce concrete with tensile elements – be it with long steel rods or short-fibered aggregates such as steel or glass fibers. As the consequence of this technology, certain process steps are required to install the reinforcement. In spite of initial approaches of adding fibers to the concrete mix as aggregates in order to enable the concrete to fully bear tensile and pressure loads, these technologies were not able to become broadly established.

Therefore, the objective of another approach is to develop fiber-reinforced concretes, which, besides general strengthening by the addition of fibers, involve installing purposefully directed fiber formations to control the strength according to defined specifications; or to weaken the product in certain areas to create articulating joints.

CONCRETE WITHOUT REINFORCEMENT

Besides controlling the reinforcement, as discussed in the previous section, another important parameter of concrete manufacturing is shaping the liquid base material so that it hardens in a specific form, typically accomplished by using formwork. The effort required for formwork can be best understood by the proportionality of the cost: in western countries, the share of cost for formwork is approximately 80 % of that for the final product. An evaluation of the embodied energy used for the overall construction reflects this fact.

Another factor is formwork induced limitations in the geometric possibilities. Common formwork systems are planar; even though research examines free-formed formwork (see chapter 3). There are approaches that try to use textile formwork to allow for geometric freedom (see chapter 4) and greater efficiency while using less material. However, with all these systems the fact remains that part of the process involves creating parts that will eventually, after the concrete part is completed, be destroyed and thus do not participate in the final construction – an insolvable problem. Therefore, the goal must be to develop a method to shape concrete that can do without formwork – and, at best, corresponds to the most efficient form possible. Technologies such as highlighted in chapters 4 and 5 related to shapeable gliding formwork or additive manufacturing methods offer potential in this respect.

For the first, an extrusion process could be controlled such that the concrete solidifies during extrusion; thus creating a linear section. Hereby it is necessary to integrate the required reinforcement on the chamber side of the matrix so that it comes to lie in the correct position for the force path. Another necessary aspect is the spatial positioning or deformation of the extruded section during the extrusion process – in order to control the cross section and to generate three-dimensional geometries.

Alternatively, additive manufacturing methods offer the potential to avoid form-giving components all together by generating the desired volume additively. In the field of plastics and metals, this development is already part of broad innovations; in the field of concrete, first approaches of development can be found under the name Contour Crafting by Prof. Behrokh Khoshnevis / California (see chapter 3). Besides this nozzle-driven method, the concrete could alternatively be deposited layer by layer which, in summation, ultimately forms the building part (see chapter 3).

If we examine such a construction in terms of the structure and the result, it is interesting that the method of additive manufacturing skips the intermediate step of wrought material that we know from processing other materials such as steel, glass, timber and masonry: the material does not need to be prefabricated into a panel or profile, cut to size, processed and positioned accurately but rather it can be immediately positioned in the desired location of the construction and is operable.

LOAD-BEARING, INSULATION, STORAGE, AND ENERGY +
Load-bearing, insulation and thermal energy storage are tasks that the material concrete is certainly able to fulfill – modern light-weight concrete and concretes with particular aggregates in particular exhibit surprisingly good insulation properties.

Against the background of the demand to improve the energetic performance of buildings, energy generating and long-term energy storage must be further developed as well. In addition to the distribution of energy (heating, cooling – see chapter 4) the same technologies can be used to generate energy; a main focus of related research. The same is true for the issue of storing energy: here, the question is whether energy storage can take place on a seasonal basis directly in the building part and without long distance losses (keyword long-term PCM storage) or by means of separate central storage away from the energy generating devices with the according storage advantages and transportation losses.

CONDITIONED CONCRETE

Picking up where we left of in the previous section, which is thinking about additional performances that might be integrated into the construction, it seems obvious to try to implement any and all physical and functional properties that a building might have, be it load-bearing structure or corner detailing. Thus, besides the already mentioned topics load-bearing, insulation and energy storage as well as those that are currently under development such as light directing (keyword "transparent concrete"), properties including the haptics of the surface (see chapter 2), extended functionality of the interior space and possibly actual transparency could be subject of future development. Another practical development might be to advance the material in a way that it can respond functionally – meaning it adapts – to changing environments and functional requirements.

Two basic approaches are conceivable: the matrix solutions on one hand, a composition consisting of individual compositions in which the individual functions are fulfilled by individual components but that is integrated in the building part as a whole. Hereby, the layering or specific arrangement of the components is of lesser importance. The components could be linked by a connecting medium, the concrete matrix, which must also enable the interaction of the components.

The alternative could be a monolithic solution; here, however, the bandwidth of functionalities is limited by the technical performance capability of the material. Thus it is of essential importance that the material itself fulfils the individual functions which are then combined in the overall construction.

AND WHAT ABOUT BEAUTY?

It goes without saying that in addition to structural and functional realization of the requirements design aspects must be considered as well. With regards to solutions in the field of concrete this means that the approaches described in this book also offer potential for new types of appearance and surface quality (see chapter 2) – a lot of aesthetic potential. But it must also be said that in the case that formative requirements / wishes cannot be realized from a technical viewpoint, designers are well capable of generating formative alternatives from existing technical solutions and to integrate them in the design. This means that not every single design requirement must be realized – design can certainly adapt to technological givens, often with exciting results!

In summarizing it can be said that the herein described developmental paths result either in complex solutions that perform highly based on specific technologies and become lighter in the sense of efficiency; however with the inherent consequence of greater effort in terms of material, construction and integration. Or solutions based on simpler technologies that deliver currently already realizable performance; naturally less efficient but more durable due to less complexity.

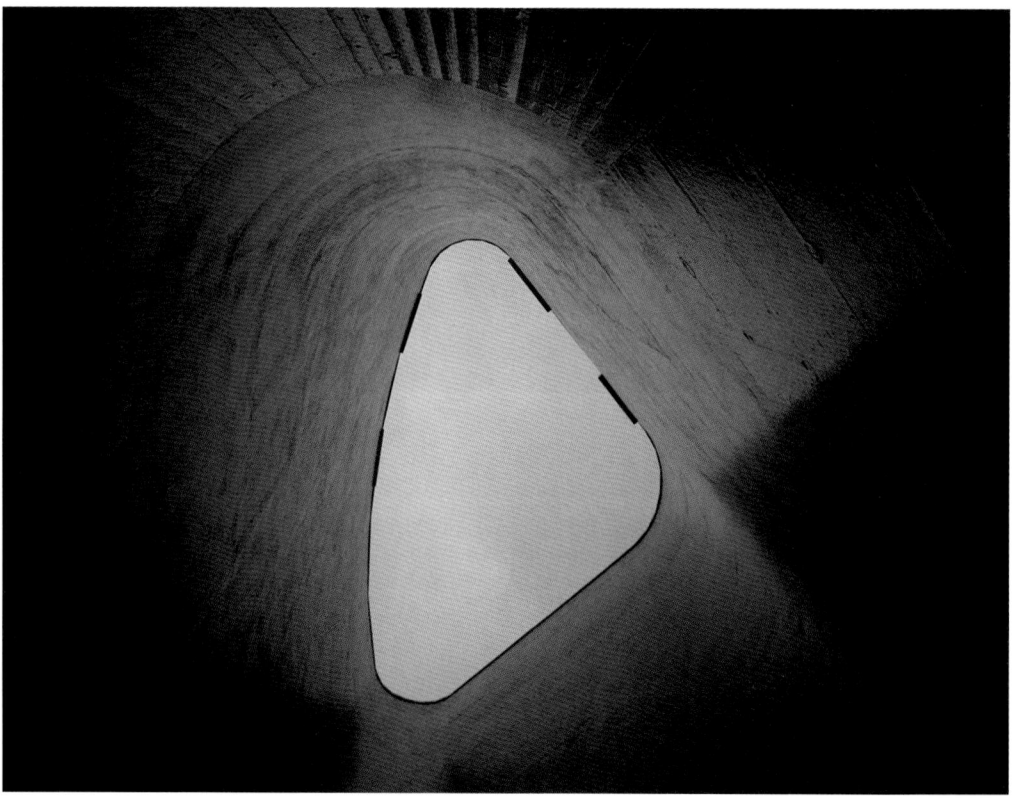

APPENDIX

CVs

ULRICH KNAACK (*1964) was trained as an architect at RWTH Aachen / Germany. After earning his degree he worked at the university as researcher in the field of structural use of glass and completed his studies with a PhD. In his professional career Knaack worked as architect and general planner in Düsseldorf / Germany, succeeding in national and international competitions. His projects include high-rise and offices buildings, commercial buildings and stadiums. In his academic career Knaack was professor for Design and Construction at the Detmold School for Architecture and Interior Architecture OWL / Germany. He also was and still is appointed professor for Design of Construction at the Delft University of Technology / Faculty of Architecture, Netherlands where he developed the Facade Research Group. In parallel he is professor for Façade Technology at the TU Darmstadt / Faculty of Civil engineering / Germany where he participates in the Institute of Structural Engineering.

He organizes interdisciplinary design workshops and symposiums in the field of facades and is author of several well-known reference books, articles and lectures.

MARCEL BILOW (*1976) studied architecture at the University of Applied Science in Detmold, completing his studies with honors in 2004. During this time, he also worked in several architectural offices, focusing on competitions and later on façade planning. Simultaneously, he and Fabian Rabsch founded the architectural office "raum204". After graduating, he worked as a lecturer and became leader of research and development at the chair Design and Constructions at Detmold School for Architecture and Interior Architecture in Detmold under the supervision of Prof. Dr. Ulrich Knaack. Since 2005, he has been a member of the Façade Research Group at the TU Delft, Faculty of Architecture and in 2012 he became head of the facade prototype course, the BuckyLab. The same year he completed his PhD thesis on climate related façades.

SASCHA HICKERT (*1976) studied architecture at Detmold School of Architecture and Interior Architecture, completing the Bachelor degree in 2010 with distinction, immediately followed by a Master Program which, in 2013 he also completed with distinction. During his studies Hickert worked as scientific assistant in different subject areas at the chair Principles of Design, Three-dimensional Design under Prof. Ernst Thevis and the chair Design and ConstructionLab. under Prof. Ulrich Knaack. During his studies he started his research activities and became a member of the ConstructionLab, Detmold. From 2011 – 2014 the research was part of the German DFG (Deutsche Forschungsgemeinschaft) SPP 1542 Light concrete construction. The topic was 'Fundamentals for the development of adaptive framework for freeform concrete structures'. Since 2012, Hickert is a member of the Façade Research Group at the TU Delft, Faculty of Architecture. His research was embedded in the teaching activities at Detmold School for Architecture and Interior Architecture (2013 – 2014). This activity is now continued at Technische Universität Darmstadt. Since 2014 he works as a scientific staff member in research and education at Technische Universität Darmstadt, Faculty of Civil Engineering, chair Facade Structures.

LINDA HILDEBRAND (*1983) completed her studies in Architecture at the Detmold School for Architecture and Interior Architecture (Germany) in 2008. Writing a diploma thesis about green certificates in the building industry, she started her career by applying the German DGNB certificate in the pilot phase. The same year she started her PhD research at TU Delft analyzing the relevance of embodied energy for the building sector as well as teaching Sustainable Construction in Detmold. She is member of TU Delft's Facade Research Group and involved in several publications such as the imagine book series and different research projects. Next to activities in research she is working for the Rotterdam office imagine envelope as consultant for ecological design. She completed her PhD in 2014 and was appointed Professor for Recyclable Construction at RWTH Aachen the same year.

TILLMANN KLEIN (*1967) studied architecture at the RWTH Aachen, completing with a degree in 1994. From then on he worked in several architecture offices, later focussing on the construction of metal and glass facades and glass roofs. Simultaneously he attended the Kunstakademie in Düsseldorf, Klasse Baukunst, completing the studies in 2000 with the title "Meisterschüler". In 1999 he was co-founder of the architecture office rheinflügel baukunst with a focus on art related projects. In 2005 he was awarded the art prize of Nordrhein-Westfalen for young artists.
Since September 2005 he leads the Facade Research Group at the TU Delft, Faculty of Architecture and since 2008 he is director of the façade consulting office Imagine Envelope b.v. in Den Haag. Tillmann Klein organises the international facade conference series

'The Future Envelope' at the TU Delft and is editor of the scientific 'Journal of Facade Design and Engineering'.

REFERENCES

Chandler, A., Pedreschi, R. (2007). Fabric
formwork. London: RIBA Publishing

Girmscheid, Kersting. (2011). Rational schalen.
Der Bauingenieur, Nr.11/2011 S. 50-51.

Heinle, Schlaich. (1996). Kuppeln aller
Zeiten – aller Kulturen. Stuttgard: Deutsche
Verlags-Anstalt.

Kind-Barkauskas et al. (2001). Beton Atlas.
Berlin: Birkhäuser Verlag.

Kowalski, R.-D.(2001). Schaltechnik im
Betonbau. Düsseldorf: Werner Verlag.

Manelius, A.M. (2012). Fabric Formwork -
Investigations into Formwork Tectonics and
stereogeneity in architectural Construction.
(Doctoral Dissertation), Royal Danish Academy
of Fine Arts.

Michel, M. (2012). Electronic Controlled
Adaptive Formwork for Freeform Concrete
Walls and Shells. In: Müller, H.S., Haist, M.,
Acosta, F. The 9th fib International PhD
Symposium in civil engineering. KIT Scientific
Publishing, Karlsruhe.

Peer, A. (1998). Der neue Zollhof in Düsseldorf.
Innovative Schalung für ein außergewöhnliches
Bauprojekt. beton 48, Nr 9, S. 538–544.

Schmitt, R. (2001) Die Schalungstechnik:
Systeme, Einsatz und Logistik. Berlin:
Ernst & Sohn.

Stacey, M. (2011)TACEY, M. (2011). Concrete: a
studio design guide. London: RIBA Publishing.

The book is made partly possible by the support of
the following partners: DFG Deutsche Forschungsgemeinschaft – SPP1542 Leicht Bauen mit Beton,
Hochschule Ostwestfalen Lippe - FB 1, Technische
Universität Darmstadt - FB 13.

With the kind support of

Hochschule Ostwestfalen-Lippe
University of Applied Sciences

TECHNISCHE
UNIVERSITÄT
DARMSTADT

ALSO PUBLISHED

- Imagine 01 – Facades
  ISBN 978-90-6450-656-7
- Imagine 02 – Deflateables
  ISBN 978-90-6450-657-4
- Imagine 03 – Performance Driven Envelope
  ISBN 978-90-6450-675-8
- Imagine 04 – Rapids
  ISBN 978-90-6450-676-5
- Imagine 05 – Energy
  ISBN 978-90-6450-761-8
- Imagine 06 – Reimagining the Envelope
  ISBN 978-90-6450-800-4
- Imagine 07 – Reimagining Housing
  ISBN 978-94-6208-036-2

CREDITS

IMAGINE 08 – CONCRETABLE

Series on technology and material development, Chair of Design of Constructions at Delft University of Technology.

Imagine provides architects and designers with ideas and new possibilities for materials, constructions and façades by employing alternative or new technologies. It covers topics geared toward technical developments, environmental needs and aesthetic possibilities.

©2015 nai010 publishers, Rotterdam
www.nai010.nl

nai010 publishers is an internationally orientated publisher specialized in developing, producing and distributing books on architecture, visual arts and related disciplines.
Available in North, South and Central America through D.A.P./Distributed Art Publishers Inc, 155 Sixth Avenue 2nd Floor, New York, NY 10013-1507, tel +1 212 627 1999, fax +1 212 627 9484, dap@dapinc.com. Available in the United Kingdom and Ireland through Art Data, 12 Bell Industrial Estate, 50 Cunnington Street, London W4 5HB, tel +44 208 747 1061, fax +44 208 742 2319, orders@artdata.co.uk

ISBN 978-94-6208-221-2

SERIES EDITORS
Ulrich Knaack, Tillman Klein, Marcel Bilow

PEER REVIEW
Prof. Dr.-Ing. Holger Techen, Frankfurt University of Applied Sciences
Prof. Dipl.-Ing. Matthias Rudolph, Staatliche Akademie der Bildenden Künste Stuttgart

AUTHORS
Ulrich Knaack, Sascha Hickert, Linda Hildebrand

CO-AUTHORS
Matthias Michel, Ahmed Hafez, Rebecca Bach

TEXT EDITING
Usch Engelmann

DESIGN
Studio Minke Themans

PRINTED BY
DeckersSnoeck, Antwerp

ILLUSTRATION CREDITS
All illustrations by the authors and people who contributed to this book, or as mentioned in the image description.